International Workshop on Itinerant-Electron Magnetism 2015

Journal of Physics: Conference Series Volume 868

Kyoto, Japan
25 – 27 September 2015

ISBN: 978-1-5108-4423-0
ISSN: 1742-6588

Printed from e-media with permission by:

Curran Associates, Inc.
57 Morehouse Lane
Red Hook, NY 12571

Some format issues inherent in the e-media version may also appear in this print version.

Printed by Curran Associates, Inc. (2017)

For permission requests, please contact the Institute of Physics
at the address below.

Institute of Physics
Dirac House, Temple Back
Bristol BS1 6BE UK

Phone: 44 1 17 929 7481
Fax: 44 1 17 920 0979

techtracking@iop.org

Additional copies of this publication are available from:

Curran Associates, Inc.
57 Morehouse Lane
Red Hook, NY 12571 USA
Phone: 845-758-0400
Fax: 845-758-2633
Email: curran@proceedings.com
Web: www.proceedings.com

Table of contents

Volume 868

International Workshop on Itinerant-Electron Magnetism
25–27 September 2015, Kyoto, Japan

Accepted papers received: 23 June 2017
Published online: 4 July 2017

Preface

Papers

IOP Conf. Series: Journal of Physics: Conf. Series **868** (2017) 011001 doi:10.1088/1742-6596/868/1/011001

Preface

In these several decades, a lot of important approaches have been conducted in order to understand the itinerant-electron magnetism. Among them epoch-making was the great success of the spin-fluctuation theory for weak itinerant ferro- and antiferromagnets by Moriya and his coworkers based upon the self-consistent renormalization of spin fluctuations since 1973 (the SCR theory of spin fluctuations). Afterwards, the spin fluctuation theory has been developed toward the unified theory between the weakly itinerant and localized moment regimes in the metallic magnets in a phenomenological way by Moriya and Takahashi (1978). Then, the SCR theory has been developed and rearranged in a quantitative way by Takahashi and Moriya in 1985, by which we can compare the experiments and the SCR theory quantitatively by means of a set of (several numbers of) spin-fluctuation parameters.

Furthermore, Takahashi has developed the spin-fluctuation theory in different approaches with some assumptions: the total spin-fluctuation amplitude conservation (TAC) and global consistency (GC) so far (1986 ~ present), which possibly leads us to the unified picture of metallic magnetism with a wide variety of itinerant-electron magnets based upon the spin-fluctuation approaches.

Since the novel superconductors with the magnetic origins have been discovered in the strongly correlated electron systems, e.g., the Heavy-Fermion compounds and intermetallics, the organic systems, the high-T_c cuprates, and Fe pnictides, moreover the electron correlations and interplays between the itinerant magnetism and the exotic superconductivity have been important, and the itinerant-electron characteristics have recently become one of the most important problems in the solid state sciences.

In organizing the International Workshop on Itinerant-Electron Magnetism, which was held during September 25-27, 2015 in the seminar house of Graduate School of Science, Kyoto University, Kyoto, Japan, we had a plan to bring together an international group of leading theoreticians and experimental scientists on magnetism to discuss advanced topics in condensed matter physics, especially related to spin fluctuations in itinerant-electron magnetism including exotic superconductivity with magnetic pairing mechanism of electron pairs, and to shape the future development of this field. We also plan to invite young scientists as well as graduate students. We hoped that such young scientists have chance to talk with invited speakers and organizers on their own interests.

Finally, this workshop is also organized in commemoration of Prof. Y. Takahashi's retirement from University of Hyogo, Japan.

(Kazuyoshi Yoshimura)

IOP Conf. Series: Journal of Physics: Conf. Series **868** (2017) 011002 doi:10.1088/1742-6596/868/1/011002

Peer review statement

All papers published in this volume of *Journal of Physics: Conference Series* have been peer reviewed through processes administered by the proceedings Editors. Reviews were conducted by expert referees to the professional and scientific standards expected of a proceedings journal published by IOP Publishing.

Theme of the Workshop on Itinerant-Electron Magnetism, and Spin Fluctuations

Kazuyoshi Yoshimura[1, 2, 3, 4]

[1]*Department of Chemistry, Graduate School of Science, Kyoto University, Kyoto 606-8502, Japan*

[2]*Research Center for Low Temperature and Material Sciences, Kyoto University, Kyoto 606-8501, Japan*

[3]*International Research Unit of Integrated Complex System Science, Kyoto University, Kyoto 606-8501, Japan*

[4]*International Research Unit of Advanced Future Studies, Kyoto University, Kyoto 606-8502, Japan*

E-mail: yoshimura.kazuyoshi.8e@kyoto-u.ac.jp, kyhv@kuchem.kyoto-u.ac.jp

Abstract: The international workshop on itinerant-electron magnetism (IWIEM) was held during September 25-27, 2015 in the seminar house of Graduate School of Science, Kyoto University, Kyoto, Japan. Here, I explain the theme of this workshop, and stress the development of itinerant-electron magnetism in several decades. The workshop was also organized in commemoration of Professor Yoshinori Takahashi's retirement from University of Hyogo, Japan. Here, I also explain some of his works contributing to the development of itinerant magnetism.

1. Itinerant-Electron Magnetism and Spin Fluctuations

In these several decades, a lot of important theoretical and experimental approaches have been conducted for understanding the itinerant-electron magnetism [1-20, and references in them]. Among them epoch-making was the great success of the spin-fluctuation theory for weak itinerant ferromagnets and antiferromagnets by Toru Moriya and his coworkers based on the self-consistent renormalization (SCR) of spin fluctuations to magnetic free energy since 1973 (the SCR theory) [6, 8-11], exceeding the Stoner mean field theory [1] and the dynamical mean field theory, called the random phase approximation (RPA) theory with no mode-mode coupling of spin fluctuations [5]. Afterwards, the spin fluctuation theory has been developed toward the unified theory between the weakly itinerant regime [6] and the localized moment regime [4] in metallic magnets by a phenomenological method by T. Moriya and Yoshinori Takahashi (1978) [8]. Furthermore, the SCR theory was successfully extended to explain the characteristic magnetic behaviors of heavy-fermion systems [17]. Then, the SCR theory has been developed and rearranged in a quantitative way by

Takahashi and Moriya in 1985 (Quantitative Aspect of Spin Fluctuations) [12], by which we can compare the experiments and the SCR theory quantitatively by means of a set of (several numbers of) spin-fluctuation parameters [12, 15, 16, 18-20].

2. SCR Theory of Spin Fluctuations

In the SCR theory of spin fluctuations for itinerant-electron magnetic systems [6, 8-14, 19, 20], the definitions of spin fluctuations, $\langle S^2\rangle = \langle S^2\rangle_T + \langle S^2\rangle_{Z.P.}$, which are magnetic excitations and band fluctuations in itinerant-electron magnetic systems, are

$$\begin{cases} \langle S^2\rangle_T = \dfrac{6}{N_0^2}\sum_q \int_0^\infty \dfrac{d\omega}{\pi} n(\omega)\,\mathrm{Im}\,\chi(q,\omega) \\ \langle S^2\rangle_{Z.P.} = \dfrac{3}{N_0^2}\sum_q \int_0^\infty \dfrac{d\omega}{\pi}\mathrm{Im}\,\chi(q,\omega) \end{cases}, \tag{1}$$

where $\langle S^2\rangle_T$ is the thermal spin fluctuations, $\langle S^2\rangle_{Z.P.}$ is the zero-point spin fluctuations, q is the wave vector of the spin fluctuation, ω is the frequency of the spin fluctuation, N_0 is the number of magnetic atoms in the system, and $\mathrm{Im}\chi(q,\omega)$ is the imaginary part of the dynamical magnetic susceptibility as a function of q and ω. Furthermore, $n(\omega)$ is the bosonic factor. The difference of $\langle S^2\rangle_T$ from $\langle S^2\rangle_{Z.P.}$ is only the presence of $n(\omega)$ [13, 19, 20]. The SCR theory of spin fluctuations, which is the mode-mode coupling theory between different wave q-vectors of spin fluctuations, has explained many experimental magnetic properties of the itinerant ferromagnets and antiferromagnets, for example, the low ferromagnetic transition temperature, called the Curie temperature T_C, and the low antiferromagnetic transition temperature, called Néel temperature T_N, in itinerant-electron systems and the dynamical measurements of spin dynamics in itinerant magnets [6, 8-20]. The Curie-Weiss law in χ above T_C was successfully explained by the T-linear increase of $\langle S^2\rangle_T$ in the SCR theory [6, 9-12, 14].

In the SCR theory of spin fluctuations, $\mathrm{Im}\chi(q,\omega)$ is usually written by the following double Lorentzian formula as [9-14, 19, 20]

$$\begin{cases} \mathrm{Im}\,\chi(q,\omega) = \dfrac{\chi(0,0)}{1+q^2/\kappa^2}\cdot\dfrac{\omega\Gamma_q}{\omega^2+\Gamma_q^2} \\ \Gamma_q = \Gamma_0 q(\kappa^2+q^2), \quad \kappa^2 = \dfrac{1}{2\bar{A}}\cdot\dfrac{N_0}{\chi} \end{cases}, \tag{2}$$

where κ and Γ_q represent the q (momentum)- and ω (energy)-widths of the spin fluctuation spectrum, and κ^2 corresponds to the inverse susceptibility, $1/\chi$. In the SCR theory, κ and Γ_q are important spin-fluctuation quantities for expressing magnetic quantities. Furthermore, Γ_0 and $\bar{A}$ are the parameters representing ω -width and q-damping width of the double Lorentzian form. Here, by utilizing a set of spin fluctuation parameters, p_s, $\overline{F_1}$, T_0 and T_A, magnetic quantities, such as T_C, and the temperature dependence of the magnetic susceptibility above T_C, can be expressed quantitatively by the SCR theory [9-16, 18-20], where p_s is the spontaneous Bohr magneton number, $\overline{F_1}$, which can be obtained experimentally from the Allott-plot (M^2 vs H/M plot (M being magnetization and H being magnetic field)), is the coefficient of the M^4- term in Landau expansion of magnetic free energy, and T_0 and T_A are the characteristic temperatures corresponding to the q- and ω - widths of the spin fluctuation spectrum deduced from Γ_0 and $\overline{A}$ as

$$\begin{cases} T_0 = \Gamma_0 q_B^3 / 2\pi \\ T_A = \overline{A} q_B^2 \end{cases}, \tag{3}$$

where q_B is the wave vector q at the Brillouin zone boundary. For example, the SCR theory gives T_C through the following SCR relation among spin-fluctuation parameters as

$$p_s^2 / 4 = \frac{15cT_0}{T_A}\left(\frac{T_C}{T_0}\right)^{4/3}, \tag{4}$$

where c is the constant being 0.3353... [12].

3. Takahashi's Theory of Spin Fluctuations

After providing quantitative evidence for the SCR theory in 1985 [12], Takahashi has developed the spin-fluctuation theory [13] in different approaches with some assumptions: total spin-fluctuation amplitude conservation (TAC) and global consistency (GC) for a few decades (1986 ~ present) [13, 19, 20], which lead us to the new unified picture of metallic magnetism with a wide variety of itinerant-electron magnets based upon the spin-fluctuation approaches [19, 20].

First, he assumed that the following TAC equation (5) is valid even in explaining magnetic properties at finite temperatures [13]. Therefore, total spin fluctuations, namely, the total square amplitude of the local spin fluctuation, are constant and conserved even in the itinerant system as [13, 19, 20]

$$\langle S_{loc}^2 \rangle = \langle S_{loc}^2 \rangle_T + \langle S_{loc}^2 \rangle_{Z.P.} = const. \tag{5}$$

Equation (5) is naturally satisfied in the localized moment system. Takahashi assumed the eq. (5) is valid even in an itinerant system, although that is not intuitive in the itinerant system. By using the relation deduced from the eq. (5), the magnetic properties at finite temperatures can be reproduced and explained [13, 15, 16, 18-23]. This allows the new unification of the itinerant-electron magnetism [13, 19, 20] from the Pauli paramagnetic weak limit [6] to the localized-moment limit, even in the metallic state [4], leading the unified relation between p_s/p_{eff} (paramagnetic Bohr magneton number obtained from the Curie constant above T_C) and T_C/T_0 independent of magnetic materials as

$$p_{eff} / p_s \approx 1.4 \times (T_C / T_0)^{-2/3}, \tag{6}$$

which gives the new universal picture (the universal p_{eff}/p_s vs T_C/T_0 plot) among various ferromagnetic materials [13, 19, 20, 22] instead of phenomenological Rhodes-Wohlfarth plot (p_{eff}/p_s vs T_C plot), which was discussed based on the mean field theory [7].

Next, Takahashi has assumed the global consistency (GC) at T_C [13]. In the Stoner and

SCR theories assumed the Landau expansion up to the M^4-term is important in magnetic free energy, there occurred the discontinuity at T_C in these theories. He has assumed that the M^6-term is important at T_C, resulting in the continuity in M at T_C (GC) [13, 19, 20]. By this assumption he obtained the following relation at T_C as

$$\left(\frac{M}{M_s}\right)^4 = 2\left[3\pi\left(2+\sqrt{5}\right)\right]^2 N_0\mu_B^2 \frac{T_C^2}{T_A^3 p_s^4}\frac{H}{M}. \tag{7}$$

which leads to the validity of M^4 vs H/M plot instead of M^2 vs H/M plot at T_C, where M_s is the spontaneous magnetization and μ_B is the Bohr magneton. Furthermore, consequently from eq. (7) we can obtain the value of T_A from the M^4 vs H/M plot experimentally. These results from Takahashi's theory of spin fluctuations have been proved quantitatively by several experiments [19-23].

4. Exotic Superconductivity and Spin Fluctuations

Since the novel superconductors have been discovered in the strongly correlated electron systems, such as heavy-fermion compounds and intermetallics [24], the organic systems [25], the high-Tc cuprates [26, 27], and Fe pnictides [28, 29], the correlations and interplays between the itinerant magnetism and the novel superconductivity, called exotic superconductivity, have been important, and the itinerant-electron characteristics have recently become one of the most difficult and important problems in the solid state sciences [30-33]. The formalism of the BCS mean field theory [34-36] should be valid even in high-T_c cuprate and iron pnictides superconductors, as well as other strongly correlated electron superconductors, such as heavy-fermion superconductors and organic superconductors, although the mediation mechanism of Cooper pairs may be different from that of BCS theory [30-33]. In high-T_c cuprates, microscopic experiments have shown that the magnetic excitations were crucial [37], leading to the possible mechanism involving magnetic interaction-mediated Cooper pairs [30-33]. In the spin-fluctuation theory, the superconducting transition temperatures T_c are found to be universally scaled by the characteristic temperature corresponding to the energy width of spin fluctuations, T_0, in exotic superconductors [32].

5. International Workshop on Itinerant-Electron Magnetism (IWIEM)

In this workshop on itinerant-electron magnetism, we had a plan to bring together an international group of leading theoreticians and experimental scientists on magnetism to discuss advanced topics in condensed matter physics, especially related to spin fluctuations in itinerant-electron magnetism including exotic superconductivity with magnetic origins, and to shape the future development of this field. We also planned to invite young scientists as well as graduate students. We hoped that such young scientists had chances to talk with invited speakers and organizers on their own interests.

Finally, this workshop was organized for a celebration of and in commemoration of Professor Yoshinori Takahashi's retirement from University of Hyogo, Japan.

6. Invited Presentations of IWIEM

The invited talks (speakers and titles) in this workshop were as follows:

Rafik Ballou, *Institute Néel, CNRS & University Joseph Fourier, Grenoble, France,* "The Spin Liquid Phase Conundrum in the Metallic Magnet YMn_2"

Nicholas Curro, *Department of Physics, University of California, Davis, California, USA,* "NMR Studies of Inhomogeneous Glassy Behavior driven by Nematic Fluctuations in Iron Arsenide Superconductors"
Minghu Fang, *Department of Physics, Zhejiang University, Hangzhou, China,* "Superconductivity and Itinerant Electron Magnetism in $TlCo_{2-x}Ni_xSe_2$ System"
Asaya Fujita, *Green Magnetic Material Research Center, AIST Chubu, Nagoya, Japan,* "Entropics: science and engineering of caloric phenomena related to itinerant-electron magnetism and spin fluctuations"
Swee K. Goh, *Department of Physics, The Chinese University of Hong Kong, Hong Kong, China,* "Structural Quantum Criticality in Superconducting $(Ca_xSr_{1-x})_3Rh_4Sn_{13}$"
Jürgen Haase, *Faculty of Physics and Earth Sciences, University of Leipzig, Leipzig, Germany,* "Uniform Spin Susceptibility and a New Phase Diagram from NMR of Cuprate Superconductors"
Zenji Hiroi, *The Institute for Solid State Physics, The University of Tokyo, Chiba, Japan,* "Revisiting the Metal-Insulator Transition of VO_2: Molecular Orbital Crystallization"
Kenji Ishida, *Department of Physics, Graduate School of Science, Kyoto University, Japan,* "Nuclear Magnetic Resonance Studies on Ferromagnetic Superconductor UCoGe"
Masayuki Itoh, *Department of Physics, Graduate School of Science, Nagoya University, Nagoya, Japan,* "Heavy-Fermion Behavior and Metal-Insulator Transition in Vanadium Oxides Investigated by NMR"
Ryosuke Kadono, *Muon Science Laboratory, Institute of Materials Structure Science, KEK, Ibaraki, Japan,* "Quasi-One-Dimensional Spin Dynamics at Low Energies in d-electron Heavy-Fermionlike Metals"
Shinsaku Kambe, *Advanced Science Research Center, Japan Atomic Energy Agency, Ibaraki, Japan,* "Distributed Twofold Ordering in URu_2Si_2"
Toru Sakai, *Graduate School of Material Science, University of Hyogo, Hyogo, Japan; Japan Atomic Energy Agency, SPring-8, Sayo, Hyogo, Japan,* "Quantum Spin Liquid in the Kagome-Lattice Antiferromagnet and Related Systems"
Noriaki Sato, *Department of Physics, Graduate School of Science, Nagoya University, Japan,* "Quantum Critical Behavior in Magnetic Quasicrystals and Approximant Crystals"
Yoshinori Takahashi, *Graduate School of Material Science, University of Hyogo, Japan,* "Theoretical Development in Itinerant Electron Ferromagnetism"
Masashi Takigawa, *The Institute for Solid State Physics, The University of Tokyo, Chiba, Japan,* "NMR Investigations on Novel Fluctuations in f Electrons"
Takeshi Waki, *Department of Materials Science and Engineering, Graduate School of Engineering, Kyoto University, Japan,* "Itinerant Electron Magnetism in Cr-based MAX Phase Compounds"
Hiroyuki Nakamura, *Department of Materials Science and Engineering, Graduate School of Engineering, Kyoto University, Japan,* "Closing Address"

Acknowledgement
The author (K.Y.) would like to thank Prof. Yoshinori Takahashi for valuable discussion. He also thanks Dr. Masaki Imai for his sincere support as a secretary of the workshop. He appreciates all the contributors to this workshop, especially proceedings' authors for their patience.

References

[1] Stoner E C "Collective Electron Ferromagnetism" 1938 *Proceedings of the Royal Society A: Mathematical, Physical and Engineering Sciences* **165** 372–414; "*Collective Electron Ferromagnetism II. Energy and Specific Heat*" 1939 *ibid.* **169** 339–371

[2] Mott N F "*The Basis of the Electron Theory of Metals, with Special Reference to the Transition Metals*" 1949 *Proc. Phys. Soc.* (London) **A62** 416-422; "*Metal-Insulator Transition*" 1968 *Rev. Mod. Phys.* **40** 677–683

[3] Hubbard J H, "*Electron Correlations in Narrow Energy Bands*" 1963 *Proc. Roy. Soc.* **A276** 238-257; 1964 *ibid.* **A277** 237-259; 1964 *ibid.* **A281** 401-419

[4] Anderson P W "*New Approach to the Theory of Superexchange Interactions*" 1959 *Phys. Rev.* **115** 2-13; "*Localized Magnetic States in Metals*" 1961 *ibid.* **124** 41-53

[5] Murata K K and Doniach S "*Theory of Magnetic Fluctuations in Itinerant Ferromagnets*" 1972 *Phys. Rev. Lett.* **29** 285–288

[6] Moriya T and Kawabata A "*Effect of Spin Fluctuations on Itinerant Electron Ferromagnetism I & II*" 1973 *J. Phys. Soc. Jpn.* **34** 639-651; 1973 *ibid.* **35** 669-676

[7] Wohlfarth E L "*Magnetic Properties of Crystalline and Amorphous Alloys: A Systematic Discussion Based on the Rhodes-Wohlfarth Plot*" 1978 J. Magn. & Magn. Mater. **7** 113-120

[8] Moriya T and Takahashi Y "*Spin Fluctuation Theory of Itinerant Electron Ferromagnetism −A Unified Picture*" 1978 *J. Phys. Soc. Jpn.* **45** 397-408

[9] Moriya T "*Recent Progress in the Theory of Itinerant Electron Magnetism*" 1979 *J. Magn. & Magn. Mater.* **14** 1-46

[10] Moriya T "*Electron correlation and magnetism in narrow-band systems*" 1981 (Springer-Verlag)

[11] Moriya T "*Spin Fluctuations in Itinerant Electron Magnetism*" 1985 (Springer-Verlag, Springer Series in Solid-State Sciences 56, Berlin Heidelberg NewYork Tokyo)

[12] Takahashi Y and Moriya T "*Quantitative Aspects of the Theory of Weak Itinerant Ferromagnetism*" 1985 *J. Phys. Soc. Jpn.* **54** 1592-1598

[13] Takahashi Y "*On the Origin of the Curie-Weiss Law of the Magnetic Susceptibility in Itinerant Electron Ferromagnetism*" 1986 *J. Phys. Soc. Jpn.* **55** 3553-3573

[14] Moriya T "*A unified picture of magnetism*" in "*Metallic Magnetism*" 1987 (ed. H. Capellmann, Springer-Verlag, Berlin)

[15] Yoshimura K, Takigawa M, Takahashi Y, Yasuoka H and Nakamura Y "*NMR Study of Weakly Itinerant Ferromagnetic* $Y(Co_{1-x}Al_x)_2$" 1987 *J. Phys. Soc. Jpn.* **56** 1138-1155

[16] Yoshimura K, Mekata M, Takigawa M, Takahashi Y and Yasuoka H "*Spin Fluctuations in* $Y(Co_{1-x}Al_x)_2$*: A Transition System from Nearly to Weakly Itinerant Ferromagnetism*" 1988 *Phys. Rev. B* **37** 3593-3602

[17] Moriya T and Takimoto T "*Anomalous Properties around Magnetic Instability in Heavy Electron Systems*"1995 *J. Phys. Soc. Jpn.* **64** 960-969

[18] Yoshimura K, Imai T, Kiyama T, Thurber K R, Hunt A W and Kosuge K "^{17}O *NMR Observation of Universal Behavior of Ferromagnetic Spin Fluctuation in the Itinerant Magnetic System* $Sr_{1-x}Ca_xRuO_3$" 1999 *Phys. Rev. Lett.* **83** 4397-4400

[19] Takahashi Y and Yoshimura K "*Itinerant Electron Magnets and Spin Fluctuations*" 2012 (1st ed., Uchida Rokakuho, ISBN4-621-07307-9)

[20] Takahashi Y "*Spin Fluctuation Theory of Itinerant Electron Magnetism*" 2013 (Springer-Verlag,

Springer Tracts in Modern Physics 253, Berlin Heidelberg)

[21] Ohta H and Yoshimura K "*Anomalous Magnetization in the Layered Itinerant Ferromagnet* LaCoAsO" 2009 *Phys. Rev. B* **79** 184407/1-5

[22] Yang J, Chen B, Wang H, Mao Q, Imai M and Yoshimura K "*Magnetic Properties in layered* ACo_2Se_2 (A=K, Rb, Cs) *with the* $ThCr_2Si_2$*-type structure*" 2013 *Phys. Rev. B* **88** 064406/1-9

[23] Imai M, Michioka C, Ohta H, Matsuo A, Kindo K, Ueda H, and Yoshimura K "*Anomalous itinerant-electron metamagnetic transition in the layered* $Sr_{1-x}Ca_xCo_2P_2$ *system*" 2014 *Phys. Rev. B* **90** 014407/1-6

[24] Steglich F, Aarts J, Bredl C D, Lieke W, Meschede D, Franz W, Schäfer H "*Superconductivity in the Presence of Strong Pauli Paramagnetism*: $CeCu_2Si_2$" 1979, *Phys. Rev. Lett.* **43** 1892–1896

[25] Lebed A G (Ed.) "*The Physics of Organic Superconductors and Conductors*" 2008 (Springer-Verlag, Springer Series in Materials Science, Vol. 110, Berlin Heidelberg New York)

[26] Bednorz J G and Müller K A "*Possible high Tc superconductivity in the* Ba–La–Cu–O *system*" 1986 *Zeitschrift für Physik B Condensed Matter* **64** 189-193

[27] Wu M K, Ashburn J R, Torng C J, Hor P H, Meng R L, Gao L, Huang Z J, Wang Y Q, Chu C W "*Superconductivity at 93 K in a New Mixed-Phase* Y-Ba-Cu-O *Compound System at Ambient Pressure*" 1987 *Phys. Rev. Lett.* **58** 908–910

[28] Kamihara Y, Hiramatsu H, Hirano M, Kawamura R, Yanagi H, Kamiya T, Hosono H "*Iron-Based Layered Superconductor*: LaOFeP" 2006 *J. Am. Chem. Soc.* **128** 10012–10013

[29] Kamihara Y, Watanabe T, Hirano M, Hosono H "*Iron-Based Layered Superconductor* $La[O_{1-x}F_x]FeAs$ ($x = 0.05$–0.12) *with* $T_c = 26$ *K*" 2008 *J. Am. Chem. Soc.* **130** 3296–3297

[30] Anderson P W "*Twenty-five Years of High-Temperature Superconductivity – A Personal Review*" 2013 *Journal of Physics: Conference Series* **449** 012001/1-10; "*Resonating Valence Bonds: A New Kind of Insulator ?*" 1973 *Mat. Res. Bull.* **8** 153-160

[31] Nagaosa N and Lee P A "*Normal-state properties of the uniform resonating-valence-bond state*" 1990 *Phys. Rev. Lett.* **64** 2450-2453; Lee P A and Nagaosa N "*Gauge theory of the normal state of high-*T_c *superconductors*" 1992 *Phys. Rev. B* **46** 5621-5639

[32] Moriya T "*Antiferromagnetic spin fluctuation and superconductivity*" 2006 *Proc. Jpn. Acad. Ser. B* **82** 1-16; Moriya T and Ueda K "*Antiferromagnetic spin fluctuation and superconductivity*" 2003 *Rep. Prog. Phys.* **66** 1299–1341; Moriya T, Takahashi Y and Ueda K "*Antiferromagnetic spin fluctuations and superconductivity in high-*T_c *oxides*" 1992 *J. Magn. & Magn. Mater.* **104-107** 456-460; Moriya T, Takahashi Y and Ueda K "*Antiferromagnetic spin fluctuations and superconductivity –A Possible Model for High* T_c *Oxides*" 1990 *J. Phys. Soc. Jpn.* **59** 2905-2915

[33] Pines D "*Finding New Superconductors: The Spin-Fluctuation Gateway to High* T_c *and Possible Room Temperature Superconductivity*" **2013** *J. Phys. Chem. B* **117** 13145–13153

[34] Bardeen J, Cooper L and Schrieffer J R "*Theory of superconductivity*" 1957 *Phys. Rev.* **108** 1175-1204

[35] Gorkov L P "*Microscopic Derivation of the Ginzburg-Landau Equations in the Theory of Superconductivity*" 1959 *Sov. Phys. JETP* **9** 1364-1367

[36] Bogoliubov N N 1958 *Nuovo Cimento* **7** 794; "*A New Method in the Theory of Superconductivity. I, II, III*" 1958 *Sov. Phys. JETP* **34** 41-55

[37] Imai T, Slichter C P, Yoshimura K and Kosuge K "*Low Frequency Spin Dynamics in Undoped and* Sr*-Doped* La_2CuO_4" 1993 *Phys. Rev. Lett.* **70** 1002-1005; Imai T, Slichter C P, Yoshimura K,

Kato M and Kosuge K "*Spin-Spin Correlation in the Quantum Critical Regime of* La_2CuO_4" 1993 *Phys. Rev. Lett.* **71** 1254-1257

Theoretical Development in Itinerant Electron Ferromagnetism

Yoshinori Takahashi

Graduate School of Material Science, University of Hyogo, Kouto 3-2-1, Kamigori, Ako, Hyogo 678-1297, JAPAN

E-mail: takahash@sci.u-hyogo.ac.jp

Abstract. In spite of the successful explanation of the Curie-Weiss law temperature dependence of the magnetic susceptibility, there exist some difficulties in the conventional spin fluctuation theory based on the expansion in powers of fluctuation amplitudes. They are now known to originate from the severe restriction on the magnetization curve. In this review, we show another approach based on the new ideas, i.e. the local spin amplitude conservation and the simultaneous treatment of temperature and external magnetic field dependences. We show various interesting magnetic properties are derived based on them and they are compared with experiments.

1. Introduction

Significant roles of spin fluctuations on various magnetic properties of itinerant electron magnets have been recognized since the publication of the paper by Moriya and Kawabata [1, 2] in 1973. The theory, now known as the SCR theory, has been successful in deriving the observed Curie-Weiss law temperature dependence of magnetic susceptibility of weak itinerant ferromagnets, such as $ZrZn_2$, Sc_3In, and MnSi. Experimentally, the presence of spin fluctuations in these magnets has been also suggested from the enhancement of the temperature linear coefficient of the specific heat at low temperatures, and the direct observation of the fluctuation amplitudes by using the polarized neutron scattering measurements. Around 1980, the theory has been regarded to be well established except for a trivial problem [3], i.e. the spontaneous magnetic moment does not vanish at the critical point with increasing temperature.

The purpose of this article is to make a brief review on the development of spin fluctuation theories, starting from the middle of 1980 in order to overcome the above difficulty of the SCR theory. It was found at that time [4] that it originates from the basic assumption of the theory, i.e. the linearity of the Arrott plot of the magnetization curve satisfied independent of temperature. At the same time, it seems better to rely on another approach, and new theory was proposed as an alternative candidate. It is based on the spin amplitude conservation and the non negligible effects of the zero-point spin fluctuations [4, 5]. As the consequence, no assumption on the magnetization curve becomes necessary between the external magnetic field H and the induced magnetization M. It was also quite successful in predicting various magnetic properties, for instance, the critical magnetic isotherm, $H/M \propto M^4$, in contradiction to the assumption of the SCR theory.

In section 2, three theories of itinerant electron magnetism and the magnetic properties predicted by them around the critical temperature are compared with each other. Section 3 is devoted to the explanation of spin fluctuation theory based on the local spin amplitude conservation. In section 4, theoretically derived magnetic properties are compared with experiments. Finally, a brief summary is presented in the last section.

2. Brief comparison of the consequences of three theories of itinerant magnetism

As with the other thermodynamic properties, magnetic properties are derived by the following free energy as a function of magnetization M and temperature T.

$$F(M,T) = F(0,T) + \frac{1}{2}a(T)M^2 + \frac{1}{4}b(T)M^4 + \frac{1}{6}c(T)M^6 + \cdots . \tag{1}$$

For instance, the magnetic isotherm, i.e. the relation between the external magnetic field H and its induced magnetization M, is given by

$$H = \frac{\partial F(M,T)}{\partial M} = a(T)M + b(T)M^3 + c(T)M^5 + \cdots . \tag{2}$$

The inverses of magnetic susceptibilities, $\chi^{-1}(T)$ and $\chi_z^{-1}(T)$, are also defined by

$$\begin{aligned} \chi^{-1}(T) &= \frac{H}{M} = a(T) + b(T)M^2 + c(T)M^4 + \cdots , \\ \chi_z^{-1}(T) &= \frac{\partial H}{\partial M} = a(T) + 3b(T)M^2 + 5c(T)M^4 + \cdots , \end{aligned} \tag{3}$$

corresponding to the applied magnetic field perpendicular and parallel direction to the static induced magnetic moment M in z-direction. The M-dependent terms in the above right hand side are neglected in the weak field limit. In the ordered phase below T_C, (3) is rewritten in the form,

$$\frac{H}{M} = b(T)\left[M^2 - M_0^2(T)\right] + \cdots , \quad M_0^2(T) = -\frac{a(T)}{b(T)}, \tag{4}$$

where $M_0(T)$ is the spontaneous magnetic moment.

In the following, we show how these magnetic properties are treated based on the three different theories of itinerant electron magnetism.

2.1. *Stoner-Wohlfarth theory*

This theory is based on the band theoretical idea. Therefore, the temperature dependences of various magnetic properties originate from that of the Fermi distribution function. According to this theory, the temperature dependence of the inverse of magnetic susceptibility, $\chi^{-1}(T)$, and the spontaneous magnetization squared, $M_0^2(T)$, are given by

$$\chi^{-1}(T) \propto (T^2 - T_C^2), \quad M_0^2(T) \propto (T_C^2 - T^2). \tag{5}$$

Around the critical temperature T_C, they are also rewritten as follows.

$$\chi^{-1}(T) \propto (T - T_C)^{\gamma}, \quad M_0^2(T) \propto (T_C - T)^{\beta}, \tag{6}$$

where $\gamma = 1$ and $\beta = 1/2$. In the same way, the magnetization curve at the critical point, $T = T_C$, is given by $H \propto M^3$, i.e. $M \propto H^{1/\delta}$ with $\delta = 3$, because $a(T_C) = 0$ is satisfied in (3). It means that the scaling law relation, $\gamma = \beta(\delta - 1)$, of the theory of phase transition is satisfied.

2.2. SCR theory of spin fluctuations

In contrast to the Stoner-Wohlfarth theory, the effect of thermal spin fluctuations is included in the free energy in this theory. Consequently, the temperature dependence of $a(T)$ in (3) is rather dominated by the effect of spin fluctuations. Therefore, the following temperature dependences of $\chi(T)$ and $M_0(T)$ have been derived around the critical point.

$$\chi^{-1}(T) \propto (T - T_C)^2, \quad M_0(T) \propto (T_C - T)^{1/2}. \tag{7}$$

Their critical indexes are given by $\beta = 2$ and $\gamma = 1/2$, respectively. On the other hand, the second coefficient $b(T)$ in the right hand side of (2) is assumed to be constant. Higher order terms than this M^3 term are also neglected in this theory. The exponent of the critical magnetic isotherm, $M \propto H^{1/\delta}$, is given by $\delta = 3$, just as in the Stoner-Wohlfarth theory. Therefore, the scaling law relation between these indexes is violated.

The above argument clearly indicates the necessity that the effect of spin fluctuations on the magnetization curve has to be taken into account. The exponent $\delta = 3$ between H and M should be modified, and rather $H \propto M^5$ is satisfied in contradiction to $b(T) > 0$ at the critical point. It also means that the linearity of the Arrott plot of the magnetization curve is not generally satisfied.

2.3. Spin fluctuation theory based on TAC and GC

This theory has been proposed to overcome the difficulty of the SCR theory. As for the temperature dependence, the same dependences as (7) derived by the SCR theory are satisfied. However, the different critical exponent $\delta = 5$ is obtained for the magnetic isotherm, $M \propto H^{1/\delta}$. The scaling law relation is therefore satisfied in this case.

In summary, both the temperature and the magnetic field dependence of spin fluctuation amplitudes have to be treated simultaneously in order to satisfy the scaling law relation. In other words, temperature dependence of the magnetization curve is not so simple, contrary to the expectations of the SCR theory.

3. Development beyond the SCR Theory

In this section, after the origin of the difficulty of the SCR theory is briefly explained, another approach based on the ideas of TAC and GC is presented.

3.1. Difficulty of the SCR theory

As shown in (2) and (3), the magnetization curve, the relation between M and H, is generally given by

$$H = a(T)M + b(T)M^3 + \cdots, \quad \therefore \ \frac{H}{M} = a(T) + b(T)M^2 + \cdots. \tag{8}$$

In terms of above equations, the assumption of the SCR theory is stated as follows.

(i) The temperature dependence of the coefficient $b(T)$ in (8) is neglected.

(ii) The higher order expansions in powers of M than those explicitly presented in (8) are assumed to be absent or neglected.

Main interest of this theory is focused on the first coefficient $a(T)$.

As will be shown later, the thermal spin fluctuation amplitudes is regarded as a function of temperature T, and the magnetic susceptibilities, i.e. $\chi(M,T)$ and $\chi_z(M,T)$ of the perpendicular and the parallel components with respect to the static magnetic moment. Around the critical point, they are dominated by the dependence proportional to $\chi^{-1/2}(M,T)$ and $\chi_z^{-1/2}(M,T)$. Then the M^2 term in $\chi_z^{-1}(M,T) \simeq \chi^{-1}(M,T) + 2bM^2 \to 2bM^2$ in the limit of $T \to T_C$, gives rise to the M linear term in the thermal fluctuation amplitude. It is known that this M-linear term is the origin of the non vanishing spontaneous magnetization at $T = T_C$. In other words, the difficulty is closely related to the temperature independent assumption of $b(T)$ in this theory.

3.2. Theory based on the spin amplitude conservation

In order to overcome the difficulties of the SCR theory, another spin fluctuation theory was proposed [4, 5]. It is based on the following two basic ideas [6].

(i) The total spin amplitude conservation (TAC)
It means that the average of the local spin amplitude squared on each magnetic site remains constant independent of temperature, and whether in the presence or absence of the external magnetic field.

(ii) The global consistency (GC) in the H-M space
No assumption is made on the magnetization curve in this theory, in contrast to (8) with T-independent $b(T)$ in the SCR theory. To find the magnetization curve rather becomes of the target of this theory. This is what the global consistency (GC) means.

Thermal average of spin amplitude squared on each magnetic atom or ion is generally written as the sum of three components,

$$\langle \mathbf{S}_{\mathrm{loc}}^2 \rangle = \sigma^2 + \langle \delta \mathbf{S}_{\mathrm{loc}}^2 \rangle_T + \langle \delta \mathbf{S}_{\mathrm{loc}}^2 \rangle_Z , \tag{9}$$

where $\mathbf{S}_{\mathrm{loc}}$ is the local spin operator in units of $2\mu_{\mathrm{B}}$ and the first term in the right hand side represents the spontaneous local spin amplitude squared. The last two terms represent the averages of thermal and zero-point fluctuation amplitudes. In the paramagnetic phase for simplicity, according to the fluctuation-dissipation theorem of statistical mechanics, the equal-time autocorrelation of the local spin amplitude of the left hand side of (9) is represented by

$$\langle \mathbf{S}_{\mathrm{loc}}^2 \rangle = \frac{3}{N_0^2} \sum_{\mathbf{q}} \int_0^\infty \frac{d\omega}{\pi} \coth(\omega/2k_{\mathrm{B}}T) \mathrm{Im}\chi(\mathbf{q},\omega). \tag{10}$$

The imaginary part of the dynamical magnetic susceptibility in units of $(2\mu_{\mathrm{B}})^2$ in the above integrand is given in the following double Lorentzian form:

$$\begin{aligned}
\mathrm{Im}\chi(\mathbf{q},\omega) &= \chi(\mathbf{q},0)\frac{\omega\Gamma_q}{\omega^2+\Gamma_q^2}, \quad \chi(\mathbf{q},0) = \chi(0,0)\frac{\kappa^2}{q^2+\kappa^2}, \\
\Gamma_q &= \Gamma_0 q(\kappa^2+q^2) = 2\pi T_0 x(y+x^2), \quad y = \frac{\kappa^2}{q_B^2}, \; x = \frac{q}{q_B}, \\
T_0 &= \Gamma_0 q_B^3/2\pi, \quad \frac{N_0}{\chi(q_B,0)} = \frac{N_0(q_{\mathrm{B}}^2+\kappa^2)}{\kappa^2\chi(0,0)} = \frac{N_0(1+y)}{y\chi(0,0)} \simeq 2T_A.
\end{aligned} \tag{11}$$

The number of magnetic atoms, the correlation wave-number squared, and the damping constant are, respectively, denoted by N_0, κ^2, and Γ_q. The zone boundary wave-number is denoted by q_B. As the measures of the distribution widths of fluctuation amplitudes in frequency and wave-vector space, two temperature scales, T_0 and T_A, are also introduced. From the decomposition, $\coth(\omega/2k_{\mathrm{B}}T) = 1+2n(\omega)$, the total amplitude in (10) is split into the thermal and the zero-point amplitudes as given by

$$\begin{aligned}
\langle \mathbf{S}_{loc}^2 \rangle_{\mathrm{T}}(y,t) &= \frac{6}{N_0^2}\sum_q \int_0^\infty \frac{d\omega}{\pi} n(\omega)\mathrm{Im}\chi(\mathbf{q},\omega) = \frac{9T_0}{T_A}A(y,T), \\
A(y,t) &= \int_0^1 dx x^3 \left[\log u - \frac{1}{2u} - \psi(u)\right], \quad u = x(y+x^2)/t, \\
\langle \mathbf{S}_{loc}^2 \rangle_{\mathrm{Z}}(y) &= \frac{3}{N_0^2}\sum_q \int_0^\infty \frac{d\omega}{\pi} \mathrm{Im}\chi(\mathbf{q},\omega) = \langle \mathbf{S}_{loc}^2 \rangle_{\mathrm{Z}}(0) - \frac{9T_0}{T_A}cy,
\end{aligned} \tag{12}$$

IOP Conf. Series: Journal of Physics: Conf. Series **868** (2017) 012002 doi:10.1088/1742-6596/868/1/012002

where $\psi(u)$ in the integrand of the above second line represents the digamma function. Both amplitudes depend on the parameter y through its dependence of $\chi(\mathbf{q}, 0)$ and Γ_q in (11). The thermal amplitude has an extra temperature dependence through the Bose distribution function $n(\omega)$.

3.3. Basic Equation for magnetic properties

The condition of the local spin amplitude conservation is regarded as the basic equation of the TAC-GC theory. If we compare the condition of (9) at any temperature and under the external magnetic field with the reference condition at the critical point with no external magnetic field, the following equation is derived by substituting (12).

$$\begin{aligned} &2A(y,t) + A(y_z,t) - c(2y+y_z) + \lambda\sigma^2 = 3A(0,t_c), \quad \lambda = \frac{T_A}{3T_0}, \\ &y_z(\sigma,t) = y(\sigma,t) + \sigma\frac{\partial y(\sigma,t)}{\partial\sigma}, \end{aligned} \tag{13}$$

The inverses of the perpendicular and the parallel components of magnetic susceptibilities are denoted by $y \propto H/M$ and $y_z \propto \partial H/\partial M$, respectively, with respect to the direction of the spontaneous moment. The spin moment per magnetic atom is denoted by σ in units of $2\mu_{\rm B}$. The important point of (13) is that it is regarded as an ordinary differential equation of $y(\sigma,t)$ as the function of σ. Therefore the σ dependence of $y(\sigma,t)$ is obtained by solving this equation. There is also no need to restrict the highest power of $y(\sigma,t)$ as a function of σ. The σ-dependence of $y(\sigma,t)$ is determined in this way by the global condition (GC) of (13) over the wide range of the variable σ.

In the absence of spontaneous or induced magnetic moment by the external magnetic field, both $y_z = y$ and $\sigma = 0$ are satisfied simultaneously. Then the temperature dependence of the magnetic susceptibility above T_c as well as the spontaneous magnetic moment below T_c are also obtained by solving the equation.

4. Magnetic Properties in More Detail

Let us show in this section, various interesting magnetic properties are derived by solving our basic equation (13). They are classified into two groups, i.e. properties that depend on external magnetic field or temperature.

4.1. Magnetic Field dependent properties

First of all, we show how magnetic isotherms in the ground state and at the critical temperature are derived as the solutions of (13). These results are also compared with experiments.

4.1.1. Magnetization curve at $T=0$ determined by Zero-Point Spin Fluctuations In the ground state, the following differential equation is satisfied, because of the absence of the thermal spin fluctuation amplitudes in (13).

$$-c(2y+y_z) + \lambda\sigma^2 = 3A(0,t_c), \quad y_z = y + \sigma\frac{\partial y}{\partial\sigma} \tag{14}$$

The σ dependence of $y(\sigma,t)$ is easily obtained by assuming

$$y(\sigma,0) = y_1(0)[\sigma^2 - \sigma_0^2(0)]. \tag{15}$$

Two parameters, $\sigma_0(0)$ and $y_1(0)$, correspond to the spontaneous magnetic moment $M_0(T)$ and the expansion coefficient $b(T)$, respectively, in (4) at $T=0$. After substitution of (15) into (14), they are determined as follows.

$$y_1(0) = \frac{\lambda}{5c} = \frac{T_A}{15cT_0}, \quad \sigma_0^2(0) = \frac{1}{cy_1(0)}A(0,t_c) = \frac{5T_0}{T_A}C_{4/3}t_c^{4/3}, \tag{16}$$

where $A(0,t_c) \simeq C_{4/3}t^{4/3}/3$, with $C_{4/3} = 1.006089\cdots$, is satisfied for small t_c. As for $\sigma_0^2(0)$, the same result has been also derived by the SCR theory.

According to the thermodynamic relation in (2), we can find the magnetic free energy $F_m(M,T)$ as the function of M. If we note the definition, $y \propto H/M$, then (15) is rewritten as the relation between H and M, as shown in the first equation of (8). By integrating the thus obtained relation of H vs M with respect to M, the following free energy $F_m(M,T)$ is derived.

$$F_m(M,T) = F_m(0,T) + N_0\left[\frac{N_0}{2\chi(T)}m^2 + \frac{F_1}{4}m^4 + \cdots\right], \quad F_1 = \frac{2T_A^2}{15cT_0}, \tag{17}$$

where $m = M/(2\mu_B N_0)$. It is very interesting that the fourth expansion coefficient F_1 with respect to m is determined by using only two spectral parameters T_0 and T_A. The reason is that the magnetic isotherm in this case is determined by the effect of zero-point spin fluctuations. On the contrary, F_1 is regarded to be determined by the density of states of conduction electrons and its derivatives at the Fermi energy in the Stoner-Wohlfarth theory and in the SCR theory.

As an application of the relation of F_1 in (18), we can estimate the values of spectral parameters T_0 and T_A, only by using the results of magnetic measurements. Note the following relations are satisfied for σ_0^2 in (16) and F_1.

$$F_1 = \frac{2T_A^2}{15cT_0}, \quad \sigma_0^2 \simeq \frac{5T_0}{T_A}C_{4/3}\left(\frac{T_C}{T_0}\right)^{4/3}. \tag{18}$$

By eliminating either T_A or T_0 from the above relations in (18), the following two results are obtained.

$$\left(\frac{T_C}{T_0}\right)^{5/6} = \frac{\sigma_0^2}{5C_{4/3}}\left(\frac{15cF_1}{2T_C}\right)^{1/2}, \quad \left(\frac{T_C}{T_A}\right)^{5/3} = \frac{\sigma_0^2}{5C_{4/3}}\left(\frac{2T_C}{15cF_1}\right)^{1/3}. \tag{19}$$

These ratios of T_0/T_C and T_A/T_C in the left hand side are, therefore, estimated from (19) because only the values of σ_0, T_C, and F_1 are included in both of the right hand sides. To check the validity of F_1 in (18), values of $4T_A^2/15T_0$ are compared in Table 1 with those of F_1 estimated from the slopes of the Arrott plots of magnetization measurements. Parameters T_0 and T_A are estimated either by inelastic neutron scattering experiments or by NMR measurements. These two values in the fourth and fifth columns seem to be in fairly good agreement with each other.

4.1.2. Critical magnetic isotherm At the critical temperature $T = T_C$, our basic equation (13) is well approximated, for $y \ll 1$, by

$$\begin{aligned} &2[A(y,t_c) - A(0,t_c)] + [A(y_z,t_c) - A(0,t_c) - c(2y + y_z) + \lambda\sigma^2 \\ &\simeq -\frac{\pi t_c}{4}(2\sqrt{y} + \sqrt{y_z}) + \lambda\sigma^2 = 0, \end{aligned} \tag{20}$$

because of the critical thermal amplitude $A(y,t)$ given by

$$A(y,t) - A(0,t) \simeq -\frac{\pi t}{4}\sqrt{y}. \tag{21}$$

Table 1. Comparison of values of $4T_A^2/15T_0$ and F_1.

Compounds	T_0(K)	T_A(K)	$4T_A^2/15T_0$(K)	F_1(K)	Ref.
MnSi	231	2.08×10^3	$5.0\ \times10^3$	9.7×10^3	neutron [7]
	171	2.11×10^3	6.94×10^3	–	NMR [8]
Ni_3Al	3590	3.09×10^4	0.71×10^5	–	neutron [9]
$Ni_{74.7}Al_{25.3}$	2860	4.05×10^4	1.53×10^5	1.0×10^5	NMR [10]
Sc_3In	565	2.00×10^5	0.66×10^5	2.0×10^5	NMR [11]
$ZrZn_2$	321	8.83×10^3	1.05×10^4	1.3×10^4	NMR [12]
$Y(Co_{1-x}Al_x)_2$					NMR [13]
$x = 0.13$	2290	1.16×10^4	1.57×10^4	2.1×10^4	
$x = 0.15$	2119	6.34×10^3	0.51×10^4	1.0×10^4	
$x = 0.17$	2093	7.03×10^3	0.63×10^4	1.6×10^4	

Both the y- and the y_z-linear terms in (20), originated from the zero-point fluctuations, are also neglected in this limit. The solution of (20) is easily obtained by assuming the σ dependence proportional to σ^4, as given by

$$y(\sigma, t_c) = y_c\sigma^4, \quad y_c = \left[\frac{4\lambda}{\pi t_c(2+\sqrt{5})}\right]^2. \tag{22}$$

The observed magnetization curve of the MnSi [14] will be a good candidate to confirm the above σ dependence. The linearity of the Arrott plot of this compound was known not to be satisfied around the critical temperature as shown in Figure 2 (right). Its origin was sometimes

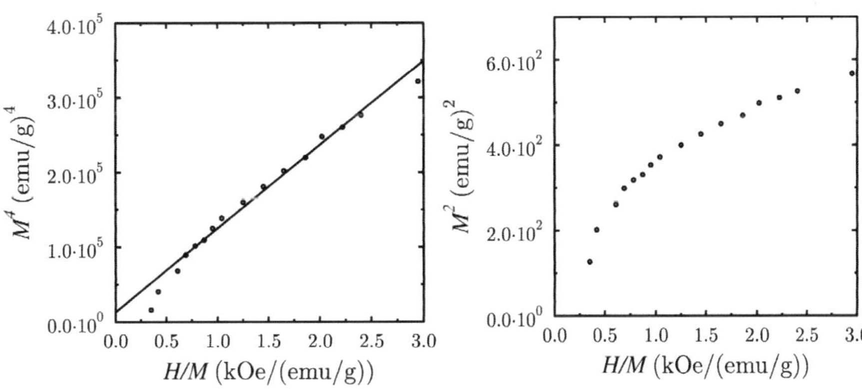

Figure 1. Magnetization curve of MnSi plotted by M^4 vs H/M (left) and M^2 vs H/M (right) at $T = 29$ K.

ascribed to the relatively larger magnitude of spontaneous magnetic moment of this compound. Therefore, it was only later that the critical M^4 vs H/M relation was found to be well satisfied as shown in the left figure. Recently good linear relations are observed for Ni [15] and the Heusler compound Co_2CrGa [16] by Nishihara *et al* , for instance, as shown in Figure 2.

Figure 2. Critical magnetic isotherms of Ni (left) and Co_2CrGa (right) by Nishihara *et al* .

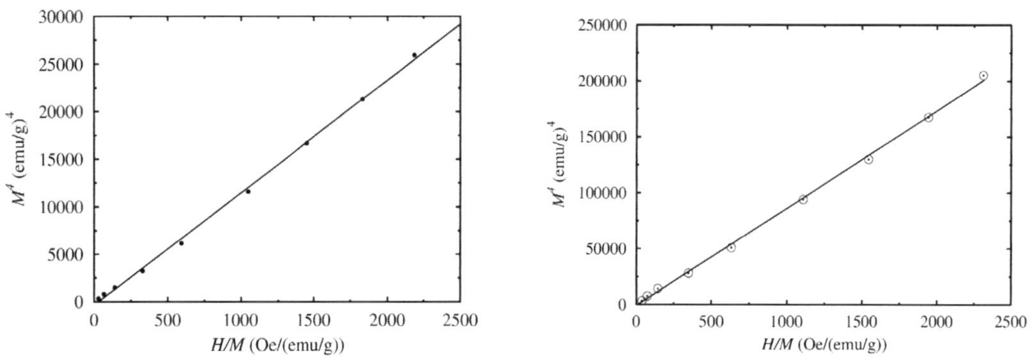

For convenience of the quantitative comparison of the theoretical prediction with experiments, the critical dependence of (22) is also written as follows.

$$\left(\frac{M}{M_g}\right)^4 = 1.20 \times 10^6 \frac{T_C^2}{w_A T_A^3 p_s^4} \frac{H}{M_g}, \tag{23}$$

where w_A is the molar weight per magnetic atom or ion, and H and M_g are measured in units of kOe and emu/g, respectively. From the experimentally observed slope of the M^4 vs H/M plot, we can estimate the value of T_A with using the values of T_C and p_s. The values of $T_A^{(c)}$ estimated in this way are shown in Table 2 for MnSi [14] and (FeCo)Si by Shimizu *et al* [17]. They are fairly in good agreement with $T_A^{(g)}$ estimated from the magnetization curve in the ground state.

Table 2. Comparison of the values of T_A estimated by (23) and those in the ground state.

Compound	$T_A^{(g)}$ (10^4 K)	$T_A^{(c)}$ (10^4 K)	Ref.
MnSi	0.218	0.129	[14]
$Fe_xCo_{1-x}Si$			[17]
$x = 0.36$	1.179	0.727	
0.48	0.998	0.727	
0.67	0.987	0.725	
0.77	1.209	0.824	
0.88	1.518	0.917	
0.91	2.273	1.268	

As the last example, the Arrott plot of numerically evaluated magnetization curves are shown in Figure 3. As a general tendency, the almost linear relations between M^2 and H/M are observed in this figure. Initial slope of curves, however, increases towards the critical temperature, and it finally diverges to infinity.

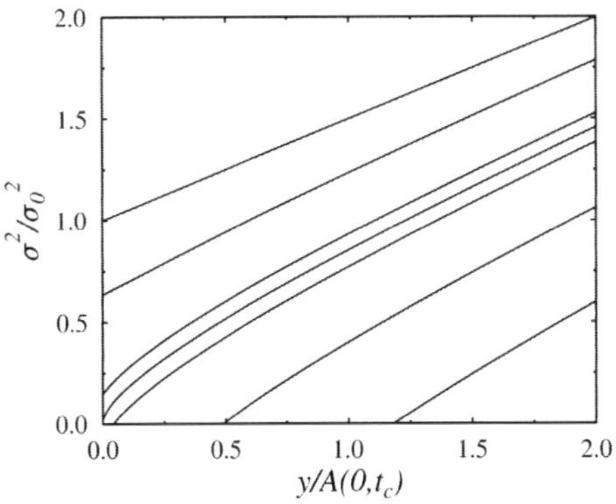

Figure 3. Arrott plot of numerically evaluated magnetization curves for $T/T_C = 0,\ 0.5,\ 0.9,\ 1,\ 1.1,\ 1.5,\ 2$ from the top.

4.2. Temperature Dependent Properties

As the typical properties of temperature dependence, the magnetic susceptibility in the paramagnetic phase and the spontaneous magnetization below the critical temperature are treated in this section.

4.2.1. Curie-Weiss law temperature dependence of magnetic susceptibility Since both of $\sigma = 0$ and $y_z = y$ are satisfied in the absence of the external magnetic field above T_C, the basic equation (13) is simply written as follows.

$$A(y,t) - c\,y = A(0,t_c). \tag{24}$$

Although it looks like the same equation of the SCR theory, the origin of the above coefficient c of the left hand side is completely different. The y-linear coefficient in this theory originates from the fourth order expansion coefficient $b(T)$ of the free energy in (1). Therefore, the almost t-linear solution of y depends on the assumption that $b(T)$ is independent of temperature. In our view, however, it depends on temperature and vanishes at the critical point, where the critical magnetic isotherm, $H \propto M^5$, is satisfied. The coefficient c in (24) rather originates from the y dependence of the amplitude of the zero-point spin fluctuations as shown in (12).

The thermal amplitude $A(y,t)$ in the left hand side of (24) increases with temperature. It is, however, compensated by the second y-linear term, to satisfy the spin amplitude conservation. The temperature dependence of the magnetic susceptibility is therefore determined as the solution $y \propto \chi^{-1}$ in (24). It is known that numerical solutions of (24) give the fairly good T-linear dependence of y, i.e. the Curie-Weiss law dependence of the magnetic susceptibility.

In the following, we show that a very interesting relation is derived from (24). If we assume the Curie-Weiss law temperature dependence of the magnetic susceptibility $\chi(T)$, its approximate T dependence is given as follows.

$$\frac{1}{N_0}\chi(T) \simeq \frac{p_{\rm eff}^2}{12(T-T_c)}, \quad \text{or} \quad \frac{N_0}{\chi(T)} \simeq \frac{12(T-T_c)}{p_{\rm eff}^2}. \tag{25}$$

From the definition of $y(t)$, the following relation is also satisfied.

$$\frac{N_0}{\chi(T)} = 2T_A y(t) \simeq 2T_A \frac{{\rm d}y(t)}{{\rm d}t}(t-t_c) = \frac{2T_A}{T_0}\frac{{\rm d}y(t)}{{\rm d}t}(T-T_c). \tag{26}$$

By equating the right hand sides of $N_0/\chi(T)$ in (25) and (26), the following relation is derived.

$$\frac{6}{p_{\rm eff}^2} \simeq \frac{T_A}{T_0}\frac{{\rm d}y(t)}{{\rm d}t} \tag{27}$$

Eliminating the ratio T_A/T_0 from (27) and the second equation of (18) for σ_0^2 by substituting (27) into (18), the following relation is finally obtained [4, 5].

$$\frac{p_{\rm eff}^2}{p_{\rm s}^2} \simeq \frac{3}{10C_{4/3}{\rm d}y/{\rm d}t}\left(\frac{T_0}{T_C}\right)^{4/3}, \tag{28}$$

where $p_{\rm s} = 2\sigma_0$ is defined. According to the distinct feature of (24), the slope dy/dt is almost independent of magnets.

Independently of the above argument, so-called Rhodes-Wohlfarth plot, i.e. $p_C/p_{\rm s}$ vs T_C plot, was already proposed [18]. The value of p_C is defined by the relation $p_C(p_C+2) = p_{\rm eff}^2$. We show in Figure 4, experimentally observed ratios of magnetic moments for compounds Y_xNi_y, plotted in two different ways against T_C or T_C/T_0. As shown in the right figure of Figure 4, almost all

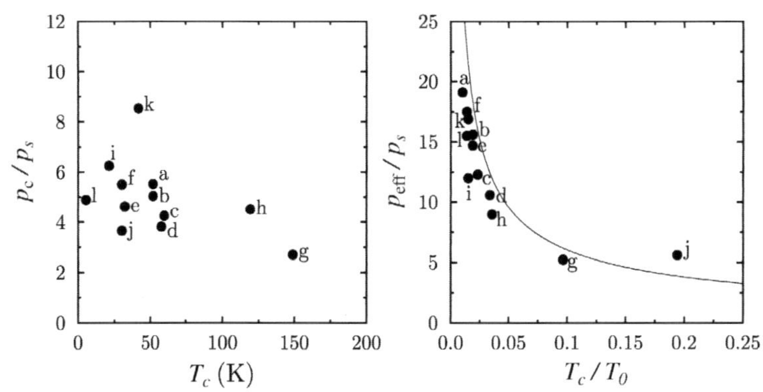

Figure 4. Rhodes-Wohlfarth plot (left) and its revised one (right) for itinerant electron ferromagnets, Y_xNi_y by Nakabayashi *et al* [19]. Solid circles of i, j, k, and l are for $ZrZn_2$, MnSi, Ni_3Al, and Sc_3In, respectively.

the ratios fall near the solid theoretical curve, confirming the validity of (28). Recently, another $\log(p_{\rm eff}/p_{\rm s})$ vs $\log(T_C/T_0)$ plot has been also proposed as shown in Figure 5.

4.2.2. Spontaneous magnetic moment In the ordered phase below T_c, the σ dependences of $y(\sigma,t)$ and $y_z(\sigma,t)$ are given by

$$y(\sigma,t) \simeq y_1(t)[\sigma^2 - \sigma_0^2(t)], \quad y_z(\sigma,t) \simeq 2y_1(t)\sigma_0^2(t) + 3y(\sigma,t), \tag{29}$$

where $\sigma_0^2(t)$ and $y_1(t)$ are the spontaneous magnetic moment squared in the absence of external magnetic field and the fourth expansion coefficient of the free energy. These parameters are determined by substituting $y(\sigma,t)$ and $y_z(\sigma,t)$ in the above (29) and σ^2 given by

$$\sigma^2 \simeq \sigma_0^2(t) + \frac{1}{y_1(t)}y(t), \tag{30}$$

into our basic equation (8). Then the following simultaneous equations for two independent parameters, $U(t) = \sigma_0^2(t)/\sigma_0^2(0)$ and $V(t) = y_z(\sigma_0(t),t)/y_z(\sigma_0(0),0)$, are obtained as the zeroth

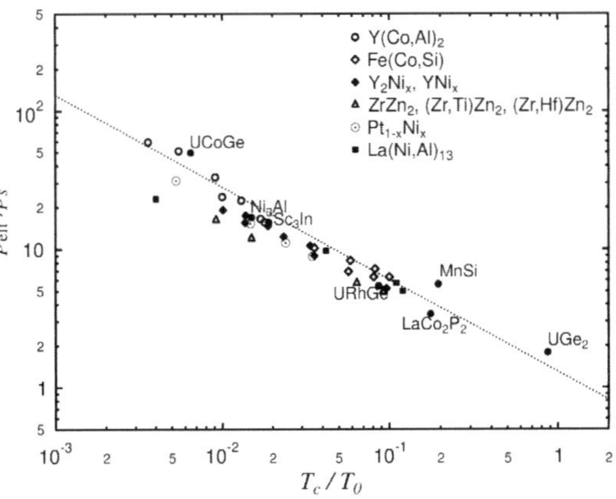

Figure 5. Deguchi-Takahashi log-log plot for various itinerant electron ferromagnets [6]. The dotted line corresponds to the relation, $p_{\rm eff}/p_{\rm s} \simeq 1.4(T_0/T_{\rm C})^{2/3}$, estimated theoretically.

and the first expansion coefficient with respect to y.

$$
\begin{aligned}
U(t) - \frac{2}{5}V(t) + \frac{1}{5A(0,t_c)}[2A(0,t) + A(y_{z0},t)] = 0,\\
V(t)\left\{1 - \frac{1}{5c}[2A'(0,t) + 3A'(y_{z0},t)]\right\} = 0.
\end{aligned}
\tag{31}
$$

Although it may seem to be a trivial problem, the y derivative of the perpendicular component of the thermal amplitude $A'(y,t)$ has to be well-defined at $y = 0$ [5] in the second equation of the above (31). The simple $\sqrt{y}$ dependence of $A(y,t)$ will lead to the divergent $A'(0,t)$. To find the t-dependence of $U(t)$ and $V(t)$ below T_C, we have to rely on some numerical calculations. As will be shown later, numerically calculated these parameters vanish simultaneously at the critical point.

Experimentally, the T^2-linear dependence of $\sigma_0^2(t)$ is observed for most of weak itinerant electron ferromagnets. On the other hand, almost no interest have been payed on the same dependence of $y_1(t)$, except for the analysis of Wohlfarth and de Chatel [20] in Figure 6, as well as those by Beille *et al* [21] for Ni-Pt alloys in Figure 7. If we confine our interest within these temperature ranges, analytical treatments become possible as will be shown below.

(i) In the limit of low temperature, reduced fourth expansion coefficient, $y_1(t)/y_1(0)$, and the spontaneous moment squared, $U(t) = \sigma_0^2(t)/\sigma_0^2(0)$, are given as follows.

$$
\begin{aligned}
\frac{y_1(t)}{y_1(0)} = \frac{V(t)}{U(t)} = 1 - \frac{c[2(\pi/2)^4 + 3]}{480A^2(0,t_c)}\left(\frac{T}{T_0}\right)^2 + \cdots \simeq 1 - \frac{b_0}{p_s^4}\left(\frac{T}{T_A}\right)^2,\\
\frac{\sigma_0^2(t)}{\sigma_0^2(0)} = U(t) = 1 - \frac{c[(\pi/2)^4 + 5(\pi/2)^2 + 4]}{360A^2(0,t_c)}\left(\frac{T}{T_0}\right)^2 + \cdots \simeq 1 - \frac{a_0}{p_s^4}\left(\frac{T}{T_A}\right)^2.
\end{aligned}
\tag{32}
$$

where $b_0 \simeq 56.91$ and $a_0 \simeq 112.1$. These relations provide us the alternative ways to estimate the parameter T_A experimentally. For example, from the T^2-linear coefficient α_0 of $U(t)$ in (33) below, the value of T_A is estimated as follows.

$$
U(t) \simeq 1 - \frac{a_0}{p_s^4}\frac{T^2}{T_A} = 1 - \alpha_0 T^2, \quad T_A = \frac{1}{p_s^2}\sqrt{\frac{a_0}{\alpha_0}}
\tag{33}
$$

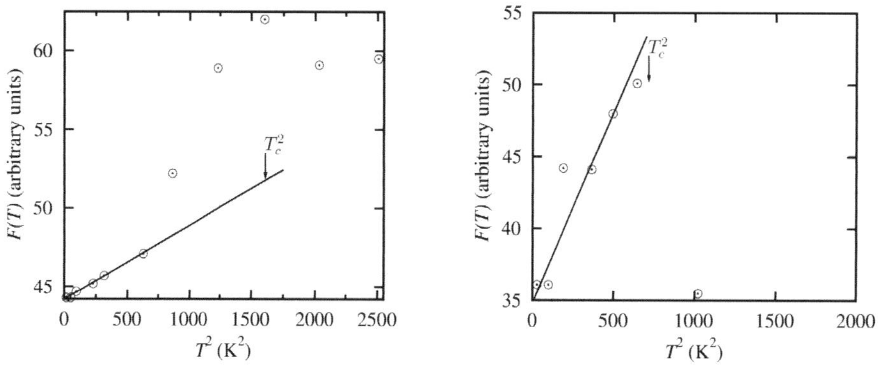

Figure 6. T^2-linear dependence of $F(T) \propto 1/y_1(t)$ for $Zr_{0.92}Ti_{0.08}Zn_2$ (left) and $ZrZn_{19}$ (right) by Wohlfarth and de Chatel (1970).

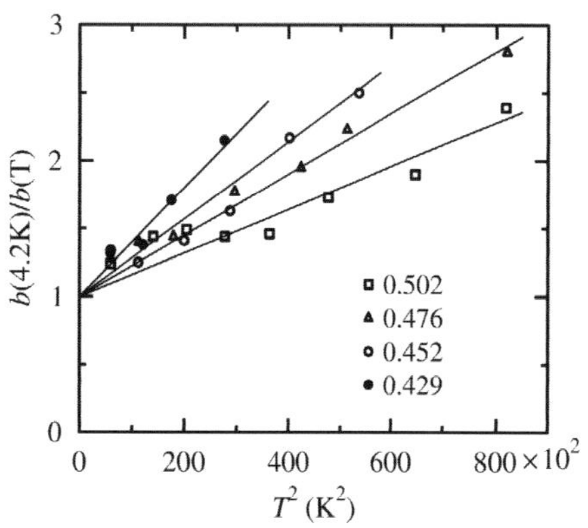

Figure 7. T^2-linear dependence of $y_1(t)$ for Ni-Pt alloys by Beille, Bloch, and Besnus (1974)

Experimentally estimated values of T_A are shown in the third column of Table 3 for several itinerant ferromagnets. They are compared well with those in the fourth column estimated from the magnetization curves in the ground state.

Table 3. Comparison of the values of T_A and $T_A^{(g)}$ estimated by two different experiments.

Compounds	α_0 (K^{-2})	T_A (K)	$T_A^{(g)}$ (K)	Ref.
$Ni_{74.7}Al_{25.3}$	2.77×10^{-3}	6.00×10^4	3.85×10^4	[22]
Ni_3Al	0.784×10^{-3}	4.51×10^4	3.09×10^4	[23]
$ZrZn_2$	2.69×10^{-3}	9.51×10^3	7.40×10^3	[24]
Y_2Ni_{15}	8.54×10^{-5}	3.41×10^4	3.51×10^4	[25]
YNi_3	1.20×10^{-3}	1.28×10^5	0.92×10^5	[26]
$Ni_{0.45}Pt_{0.55}$		1.0×10^4	0.69×10^4	[27]

(ii) Around the critical temperature below T_C, the following temperature dependence of $U(t)$ and $V(t)/U(t)$ are derived from (31).

$$\begin{aligned} U(t) &= \frac{\sigma_0^2(t)}{\sigma_0^2(0)} \simeq a_c[1-(T/T_C)^{4/3}], \\ \frac{V(t)}{U(t)} &= \frac{y_1(t)}{y_1(0)} \simeq b_c[1-(T/T_C)^{4/3}], \end{aligned} \tag{34}$$

where $a_c = 7/5$ and $b_c \simeq \dfrac{640cC_{4/3}}{21\pi^2 t_c^{2/3}} \gg 1$ for $t_c \ll 1$.
Finally, we show in Figure 8, numerically calculated results of the temperature dependence of $\sigma_0^2(t)/\sigma_0^2(0)$ and $y_1(t)/y_1(0)$ below T_C. The steep decrease of the dashed curve of

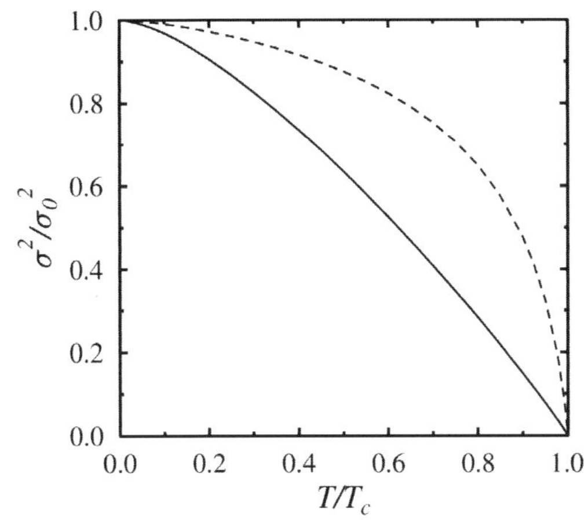

Figure 8. Numerically evaluated spontaneous magnetic moment σ^2/σ_0^2 (solid line) and the fourth expansion coefficient $y_1(t)/y_1(0)$ of the free energy (dashed line).

$y_1(t)/y_1(0)$ around the critical point results from the $1/t_c^{2/3}$ dependence of the coefficient b_c in (34).

5. Summary

Since the successful explanation of the Curie-Weiss law dependence of the magnetic susceptibility, several misleading statements have been made in the SCR theory. They are closely related to its basic concepts. Its strict condition on the magnetization curve has actually caused the problems around the critical point. The purpose of this review is to clarify the origin of these problems. On the other hand, we show how many interesting magnetic properties are derived from another theory of spin fluctuations based on the quite different basic ideas of TAC and GC.

To conclude, magnetic properties of itinerant electron ferromagnets should be understood according to the following basis.

- Spin amplitude conservation plays predominant rolls in our understanding of itinerant electron magnets.
  It implicitly implies the significant role of zero-point spin fluctuations.
- The linearity of the magnetization curve between H/M and M^2 is not generally satisfied. At the critical temperature, for instance, M^4 vs H/M relation is rather satisfied.

- Temperature and external field dependence of itinerant electron magnets are not so simple. They will change simultaneously with temperature or under the influence of the external magnetic field, contrary to the expectation of early studies.

Finally, as applications of the spin fluctuation theory presented in this review, the effects of spin fluctuations on the specific heat and the magneto-volume effects have also been treated [6, 28, 29]. Both dependences on temperature and external magnetic field are treated in consistent with thermodynamic relations.

Acknowledgments

The author thanks Prof. K. Yoshimura, Prof. H. Nakamura, Prof. N. Sato, Prof. H. Kobayashi, Dr. K. Shimizu, Dr. R. Nakabayashi, Prof. Y. Tazuke, Prof. S. Murayama, Prof. M. Shiga, Prof. T. Kanomata, Prof. H. Nishihara, Prof. T. Goto, Prof. K. Koyama, Prof. K. Fukamichi, Dr. A. Fujita, Prof. K. Deguchi, Dr. T. Koyama for valuable discussion on experiments.

References

[1] Moriya T, Kawabata A 1973 *J. Phys. Soc. Japan* **34** 639
[2] Moriya T, Kawabata A 1973 *J. Phys. Soc. Japan* **35** 669
[3] Moriya T 1985 *Spin Fluctuations in Itinerant Electron Magnetism* (Berlin Heidelberg: Springer)
[4] Takahashi Y 1986 *J. Phys. Soc. Japan* **55** 3553
[5] Takahashi Y 2001 *J. Phys.: Condens. Matter* **13** 6323
[6] Takahashi Y 2013 *Spin Fluctuation Theory of Itinerant Electron Magnetism* (Heidelberg: Springer)
[7] Ishikawa Y, Noda Y, Uemura Y J, Majkrzak C F, Shirane G 1985 *Phys. Rev. B* **31** 5884
[8] Yasuoka H, Jaccarino V, Sherwood R C, Wernick *J. Phys. Soc. Japan* **44** 1978
[9] Bernhoeft N R, Cole I, Lonzarich G G, Mitchell P W, Paul D McK 1982 *Phys. Rev. B* **28** 422
[10] Umemura T, Masuda Y 1983 *J. Phys. Soc. Japan* **52** 1439
[11] Hioki T, Masuda Y 1977 *J. Phys. Soc. Japan* **43** 1200
[12] Kontani M 1977 *J. Phys. Soc. Japan* **42** 83
[13] Yoshimura K, Takigawa M, Takahashi Y, Yasuoka H, Nakamura Y 1987 *J. Phys. Soc. Japan* **56** 1138
[14] Bloch D, Voiron J, Jaccarino, V and Wernick J H 1975 *Phys. Lett.* A **51** 259
[15] Nishihara H, Komiyama K, Oguro I, Kanomata T and Chernenko V 2007 *J. Alloys Compd.* **442** 191
[16] Nishihara H, Furutani Y, Wada T, Kanomata T, Kobayashi K, Kainuma R, Ishida K and Yamauchi T 2010 *J. Phys: Conf. Ser.* **200** 032053
[17] Shimizu K, Maruyama H, Yamazaki H, Watanabe H 1990 *J. Phys. Soc. Japan* **59** 305
[18] Rhodes P R and Wohlfarth E P 1963 *Proc. R. Soc. (London)* **273** 247
[19] Nakabayashi R, Tazuke Y, Murayama S 1992 *J. Phys. Soc. Japan* **61** 774
[20] Wohlfarth E P and de Chatel P F 1970 *Physica* **48** 477
[21] Beille J, Bloch D, Besnus M J 1974 *J. Phys.F: Met. Phys.* **4** 1275
[22] Sasakura H, Suzuki K, Masuda Y 1984 *J. Phys. Soc. Japan* **53** 754
[23] de Boer F R, Schinkel C J, Biesterbos J, Proost S 1969 *J. Appl. Phys.* **40** 1049
[24] Ogawa S 1972 *Tech. Rep. 735, Electrotechnical Lab.*
[25] Gignoux D, Lemaire R, Molho P 1980 *J. Magn. Magn. Mater.* **21** 119
[26] Gignoux D, Lemaire R, Molho P, Tasset F 1980 *J. Magn. Magn. Mater.* **21** 307
[27] Koyama K, Sasaki H, Kanomata T, Watanabe K, Motokawa M 2003 *J. Phys. Soc. Japan* **72** 767
[28] Takahashi Y, Nakano H 2004 *J. Phys.: Condens. Matter* **16** 4505
[29] Takahashi Y, Nakano H 2006 *J. Phys.: Condens. Matter* **18** 521

IOP Conf. Series: Journal of Physics: Conf. Series **868** (2017) 012003 doi:10.1088/1742-6596/868/1/012003

Nature of distributed two-fold ordering in URu_2Si_2

S. Kambe[1], Y. Tokunaga[1], H. Sakai[1], T. Hattori[1], & R.E. Walstedt[2]
[1]Advanced Science Research Center, Japan Atomic Energy Agency, Tokai-mura, Ibaraki 319-1195, Japan
[2]Physics Department, The University of Michigan, Ann Arbor, MI 48109, USA
E-mail: kambe.shinsaku@jaea.go.jp

Abstract. In previous ^{29}Si NMR basal–plane field rotation measurements, a weak component of two-fold broadening has been identified in the hidden–order (HO) state of URu_2Si_2. In addition, it has been revealed that the amplitude of two-fold ordering is distributed, a possible connection with imperfections in the single–crystal specimen used. In the present report, possible mechanisms for distributed two-fold ordering are discussed. As candidates, the effects of imperfections on the dotriacontapolar and hexadecapolar ordered states are considered.

1. Introduction

The nature of the second order phase transition at $T_0 \sim 17.5$ K in the heavy fermion compound URu_2Si_2 remains unclear up to now[1, 2, 3]. Since the order parameter of this transition has not ever been clearly identified, such order has been termed "hidden order" (HO) [4]. Owing to the tetragonal crystal structure of URu_2Si_2 (space group $I4/mmm$) [1], the physical properties of the disordered, *i.e.* paramagnetic, state of this compound exhibit fourfold rotational symmetry in the basal (001) plane. Based on in-plane anisotropy measurements of the torque magnetic susceptibility and on cyclotron resonance data, the fourfold symmetry has been reported to be spontaneously broken below T_0, where a twofold rotationally symmetric state has been found to appear [5, 6]. Elastoresistance measurements also indicate an orthorhombic distortion along the [110] direction at T_0 [7]. As a consequence of domain formation, the magnitude of the twofold symmetric ordering depends strongly on sample size and can only be detected in susceptibility measurements on very small (nearly mono–domain) samples [5]. On the other hand, orthorhombic distortion appears only in high purity samples with very low residual resistivity (relative resistance ratio between 300 K and 2 K: RRR $\sim$ 670), indicating that the twofold anisotropy seems to depend not only on sample size but also on sample quality [8].

In a previous report [9], the magnitude of twofold susceptibility anisotropy below T_0 in URu_2Si_2 has been estimated by means of ^{29}Si NMR line broadening, which does not depend explicitly on sample size. From the latter analysis the degree of twofold anisotropy in a sample with RRR$\sim$ 70 was estimated to be 15 times smaller than that determined via static susceptibility measurements on a much smaller sample [5]. Subsequently, a detailed analysis of the line broadening [10] showed the two-fold hyperfine field amplitudes to be widely distributed. In addition, two-fold hyperfine field amplitudes appeared to be distributed around the NMR shift of the four-fold state. It may be possible that the two-fold nature could even disappear in a purer sample. Recently, we have found that the linewidth results can also be reproduced with an RKKY broadening model if two–fold magnetic impurity centers are distributed at $0.1 \sim 1.0\%$

of U sites [11]. In this description, over ~99% of U sites have four-fold symmetry. This may not be surprising, since the previously observed two-fold behavior did not appear to reflect the principal order parameter. Thus, the magnitude of it depends on sample quality, whereas the specific heat anomaly at the hidden order transition is rather insensitive indicating that the hidden order itself is well developed in lower–quality samples.

In this report, we probe the origins of distributed two-fold hyperfine fields. Specifically, we consider possible two-fold states induced by crystal imperfections in cases of multipolar ordering such as dotriacontapole or hexadecapole states.

2. Experimental

Experimental details have been presented and discussed in a previous report [9]. Single crystals were grown using the Czochralski method in a tetra-arc furnace under an argon gas atmosphere, as previously described [12]. A single crystal sample with a nearly perfect cylindrical shape (1 mm $\phi \perp$ [001]×3 mm$\|$ [001]), with narrow facets on the (110) planes, was used. Since the cross-section of the sample through the basal plane is nearly a perfect circle, no demagnetization effects need to be considered. High sample purity was confirmed by a residual resistivity of ~ 5 $\mu\Omega$cm (RRR ~ 70). The resistivity was measured on a piece cut ($2 \times 0.5 \times 0.5$ mm^3) from the same single crystal used for NMR measurements. As pointed out previously [12], there is a distribution of RRR values in a large single crystal. As a consequence the measured value 70 should be regarded as an average RRR value for the NMR sample. It should be noted that URu_2Si_2 single crystals with RRR> 50 are already high quality single crystals in the usual sense [12], thus X-ray Laue diffraction and EPMA measurements cannot be used for evaluating the defect density in single crystal samples with RRR> 50.

Measured on the same piece as used for the resistivity, the T-dependence of the static magnetic susceptibility χ is almost identical with that of a very high quality sample with RRR> 600. No trace of ferromagnetic or antiferromagnetic long-range order was detected down to 2 K.

The natural abundance of the NMR isotope ^{29}Si (I = 1/2, gyromagnetic ratio $\gamma_{29}/2\pi$=0.84577 kHz/Oe) is only 4.7%. This has prevented highly accurate Si NMR measurements in URu_2Si_2 up to now. For the present study, a single crystal sample with a 53% enriched ^{29}Si isotope has been prepared, improving the NMR sensitivity by a factor ~ 11. As there are no quadrupolar interactions for $I = 1/2$ nuclei, ^{29}Si NMR spectra reflect only magnetic shift and broadening effects.

A standard pulsed NMR spectrometer was used, with magnetic field provided by a 12 T superconducting magnet in the persistent mode. NMR spectra were obtained using Fast-Fourier-Transformation (FFT) of spin-echo signals at a fixed applied magnetic field.

3. Results and Discussion

The two-fold amplitude can be defined in terms of the Knight shift anisotropy $K_{nem} \equiv K_{H\|[110]} - K_{H\|[1\bar{1}0]}$ [9], where $K_{nem} = 0$ in the four-fold state. In the case of two domains, the NMR spectra would be a superposition of two Lorentzians, one from each domain, with $K_{nem} = const > 0$. However, such a superposition cannot reproduce completely the observed spectra. Instead, the observed NMR spectra can be fitted better if a distribution of K_{nem} values is assumed in each domain [10]. As shown in Fig. 2, the no-distribution case shows a deviation between the calculation and the experiment around the peak, while the distributed case can reproduce it. For the distributed case, the distribution $g(K_{nem})$ of two-fold amplitudes K_{nem} is assumed to be [10]:

$$g(K_{nem}) = \frac{\Delta K_{nem}}{\pi(K_{nem}^2 + \Delta K_{nem}^2)}, \tag{1}$$

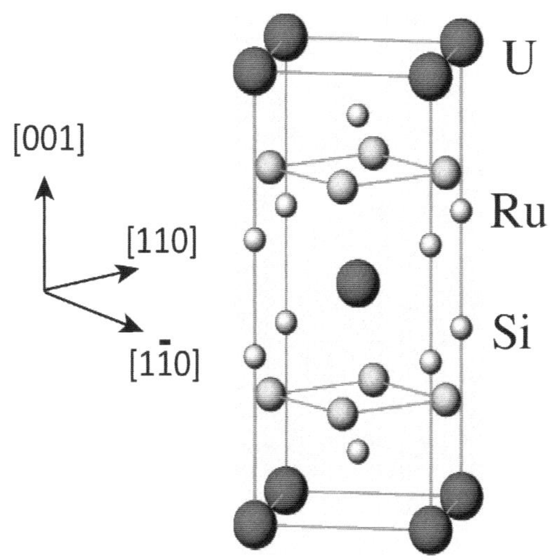

Figure 1. The crystal structure ($I4/mmm$) of URu_2Si_2. The local symmetry of the Si is tetragonal $4mm$.

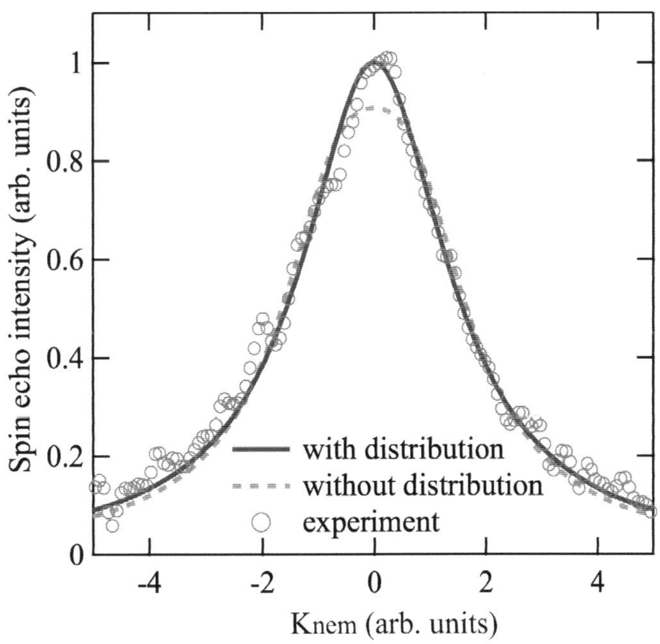

Figure 2. ^{29}Si NMR spectra with H = 5.19 T along the [110] axis at 5 K in the hidden order state of URu_2Si_2. Solid line is obtained using a distributed two-fold amplitude (Eq. 1) for a two-domain state. Dashed line is obtained using non-distributed (uniform) two-fold amplitude for a two-domain state.

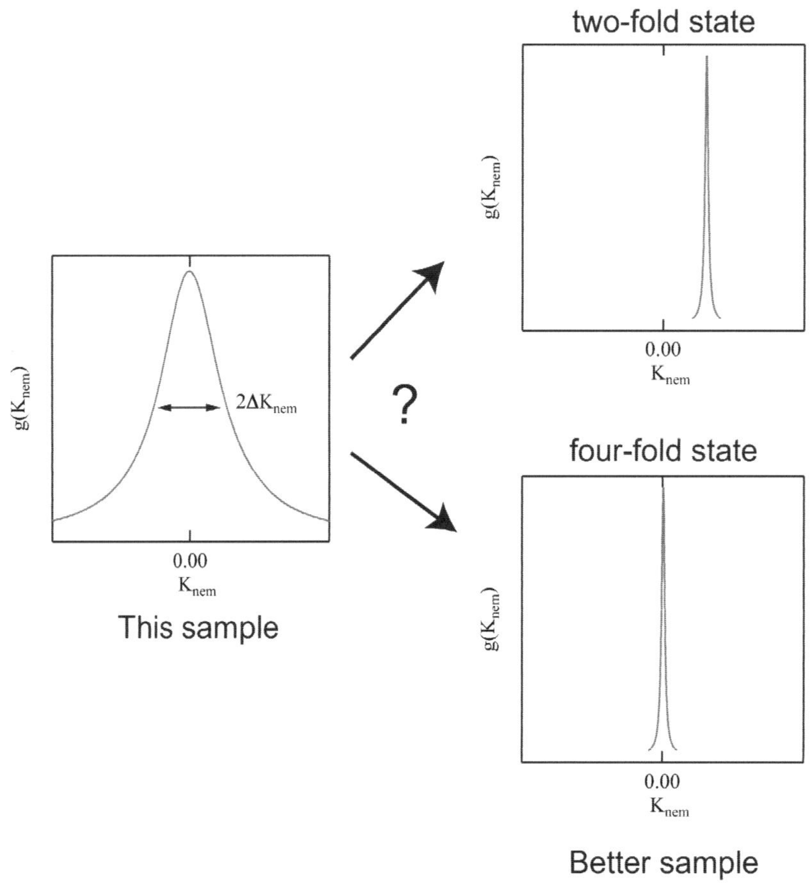

Figure 3. In the present sample, it seems that the two-fold amplitude K_{nem} is distributed around zero with the width ΔK_{nem} (left figure). In a better sample, ΔK_{nem} would be smaller (right figures). However, it is not clear whether the center position would be $K_{nem} = 0$ (four-fold state) or $K_{nem} > 0$ (two-fold state) in a better sample up to now.

where ΔK_{nem} is the distribution range of two-fold amplitudes. Also, $g(K_{nem})$ seems to have a peak at $K_{nem} = 0$ in the present sample. If ΔK_{nem} becomes zero under this situation in a better sample, there would be no two-fold amplitude, corresponding to a four-fold ordered state. At the present stage, we note that it is not clear whether the peak would locate at $K_{nem} = 0$ (four-fold state) or $K_{nem} > 0$ (two-fold state) in a better sample that exhibits intrinsic properties. In actual fact, a previous study implied that the two-fold state is prominent in a better sample [8].

Next, we discuss possible mechanisms for distributed two-fold hyperfine fields. Such hyperfine fields may appear perhaps due to distributions of impurities, dislocations or local stress (IDLS), *i.e.* two-fold states would be induced when the hidden order is broken by IDLS. In the four-fold paramagnetic state, such a two-fold state due to IDLS has not been identified [9], thus such a two-fold state may reflect the hidden order. Recently, dotriacontapolar [13] and hexadecapolar [14] order have been proposed as candidates for the hidden order parameter. These two possibilities are discussed below.

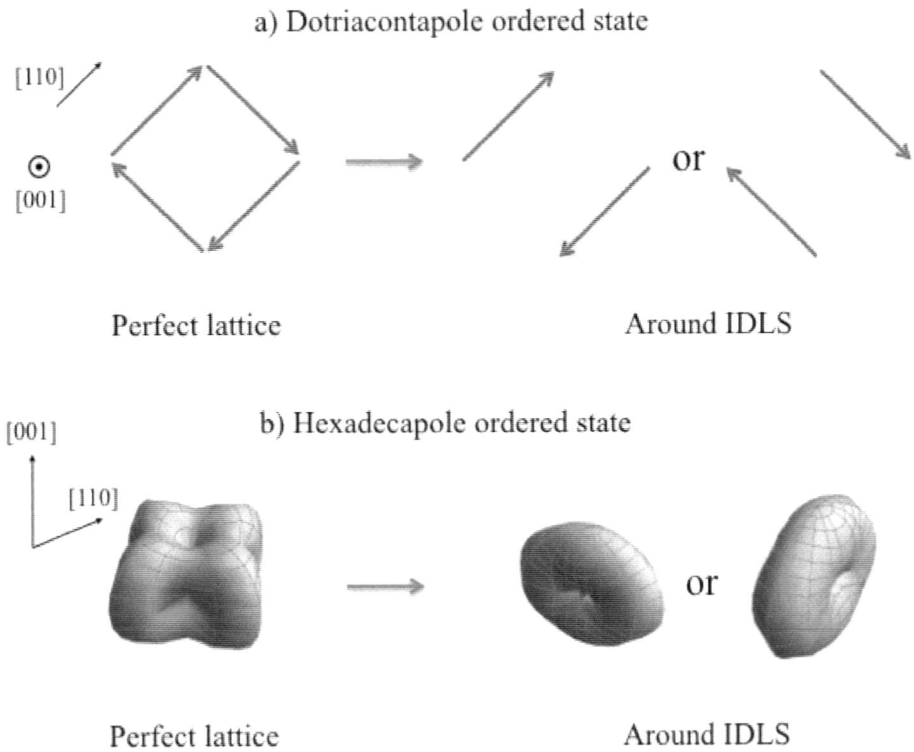

Figure 4. Schematic view of impurities, dislocations and local stress (IDLS) effects. a) Dotriacontapole case: a small moment may appear at the U site along the [110], [$1\bar{1}0$], [$\bar{1}10$] or [$\bar{1}\bar{1}0$] directions with the same population at the U site in a perfect lattice (left). Although the two-fold state appears at U sites locally, this state is effectively four-fold. Consequent to IDLS, two types of two-fold sublattices could appear, which could generate two domains (right). b) Hexadecapole case: this shape represents the density of electrons around the U site. The order parameter is locally four-fold in the perfect lattice (left). Owing to IDLS, two type of quadrupolar ordering could appear which could cause the two domains (right). As an example, O_{22}-type quadrupolar order is presented here.

3.1. Dotriacontapole scenario

In the dotriacontapolar state, a tiny pseudo-spin moment is induced at the U site along the [110] direction antiferromagnetically [13]. In this scenario, a two-fold state appears locally; however, owing to degenerate states with ordered moments along [110], [$1\bar{1}0$], [$\bar{1}10$] and [$\bar{1}\bar{1}0$] at the U site with equal populations, the NMR spectra keep four-fold symmetry. Through IDLS this equilibrium could be broken, inducing a preference for certain sublattices which show a two-fold symmetry [Fig. 4a)]. In addition, appearance of local antiferromagnetic dipolar ordering through IDSL would be possible, since the energy levels of dotriacontapolar and dipolar ordered states are quite similar and can easily be reversed by *e.g.* pressure [13, 15]. In a perfect crystal, if a mono–domain can remain dominant, the two-fold state case in Fig. 3 will be observed, while a four-fold case will be observed if all domains have the same population. Therefore, a two-fold amplitude may depend strongly on sample conditions. Considering that ordered moments perpendicular to an applied field may not affect K_{nem}, both states in Fig. 3 could be realized for the dotriacontapole case.

3.2. Hexadecapole scenario

In the hexadecapole state, four-fold symmetry is locally preserved at the U site [14]. However, when uniaxial stress along the [110] or [100] directions is applied, the two-fold quadrupolar states O_{xy} and $O_{x^2-y^2}$, respectively, could be induced [14]. Thus, it is expected that similar behavior would occur around impurities and dislocations [Fig. 4b)]. If a finite $K_{nem} > 0$ were intrinsic (the two-fold state case in Fig. 3), then the hexadecapole scenario would not occur. Thus, in a better sample the four-fold case in Fig. 3 would be observed for the hexadecapole scenario.

It should be noted that the detailed mechanism for a Lorentzian–like distribution of two-fold hyperfine fields is not understood at present. The above scenarios just indicate possible origins for such a distribution. The two-fold amplitude may well depend on the type of ordering and on imperfections. Recently, we have found that two-fold magnetic centers at $0.1 \sim 1\%$ of U sites could reproduce the NMR data based on an RKKY transferred HF coupling model [11]. The proposed two-fold states in Fig. 4 could correspond to these magnetic centers. A crucial difference between the dotriacontapole and hexadecapole ordering is that time–reversal symmetry is broken in the dotriacontapole state, whereas it is not broken in the hexadecapole state. Up to now, however, it is still not clear whether time reversal symmetry is broken or not in the HO state of URu_2Si_2.

4. Conclusion

Based on a detailed analysis of ^{29}Si NMR linewidths in the hidden order state of URu_2Si_2, the two-fold hyperfine fields at Si sites are found to be distributed. The effect of crystalline imperfections on dotriacontapolar and hexadecapolar multipolar ordering has been discussed as the possible origin of distributed two-fold hyperfine fields. In practice, we suggest that a two-fold state could appear around the imperfections in these states. To distinguish between these two types of ordering, sample quality dependence of the two-fold hyperfine fields as well as determination of time reversal symmetry would be useful.

Acknowledgments

The high quality sample for our measurements was prepared by T.D. Matsuda, Y. Haga and Z. Fisk, which is greatly appreciated. We thank K. Kubo for preparing the figures of hexadecapole and quadrupole ordering in Fig. 4b). We are grateful for stimulating discussions with T. Shibauchi, H. Ikeda, M.-T. Suzuki, Y. Matsuda, D. Aoki, F. Bourdarot, J. A. Mydosh, H. Amitsuka and J. Flouquet. A part of this work was supported by a Grant-in-Aid for Scientific Research on Innovative Areas Heavy Electrons (No. 20102006 and No. 20102007) by the MEXT, and the REIMEI Research Program of JAEA.

References

[1] Palstra T T M, Menovsky A A, van den Berg J, Dirkmaat A J, Kes P H, Nieuwenhuys G J and Mydosh J A 1985 *Phys. Rev. Lett.* **55** 2727.
[2] Schlabitz W, Baumann J, Pollit B, Rauchschwalbe U, Mayer H M, Ahlheim U and Bredl C D *Z. Phys. B Condensed Matter* **62** 171.

[3] Maple M B, Chen J W, Dalichaouch Y, Kohara T, Rossel C, Torikachvili M S, McElfresh M W and Thompson J D 1986 *Phys. Rev. Lett.* **56** 185.
[4] Tripathi V, Chandra P and Coleman P 2007 *Nature Phys.* **3** 78.
[5] Okazaki R, Shibauchi T, Shi H J, Haga Y, Matsuda T D, Yamamoto E, Onuki Y, Ikeda H and Matsuda Y 2011 *Science* **331** 439.
[6] Tonegawa S, Hashimoto K, Ikada K, Lin Y-H, Shishido H, Haga Y, Matsuda T D, Yamamoto E, Onuki Y, Ikeda H, Matsuda Y and Shibauchi T 2012 *Phys. Rev. Lett.* **109** 036401.
[7] Riggs S C, Shapiro M C, Maharaj A V, Raghu S, Bauer E D, Baumbach R E, Giraldo-Gallo P, Wartenbe M and Fisher I R 2015 *Nature Comms.* DOI: 10.1038/ncomms7425.
[8] Tonwgawa S, Kasahara S, Fukuda T, Sugimoto K, Yasuda N, Tsuruhara Y, Watanabe D, Mizukami Y, Haga Y, Matsuda T D, Yamamoto E, Onuki Y, Ikeda H, Matsuda Y and Shibauchi T 2014 *Nature Comms.* DOI: 10.1038/ncomms5188.
[9] Kambe S, Tokunaga Y, Sakai H, Matsuda T D, Haga Y, Fisk Z and Walstedt R E 2013 *Phys. Rev. Lett.* **110** 246406. See also the supplement for this paper.
[10] Kambe S, Tokunaga Y, Sakai H and Walstedt R E 2015 *Phys. Rev. B* **91** 035111.
[11] Walstedt R E, Kambe S, Tokunaga Y and Sakai H 2016 *Phys. Rev. B* **93** 045122.
[12] Matsuda T D, Hassinger E, Aoki D, Taufour V, Knebel G, Tateiwa N, Yamamoto E, Haga Y, Onuki Y, Fisk Z and Flouquet J 2011 *J. Phys. Soc. Jpn.* **80** 114710.
[13] Ikeda H, Suzuki M T, Arita R, Takimoto T, Shibauchi T and Matsuda Y 2012 *Nature Phys.* **8** 528.
[14] Kusunose H and Harima H 2011 *J. Phys. Soc. Jpn.* **80** 084702.
[15] Hassinger E, Knebel G, Izawa K, Lejay P, Salce B and Flouquet J 2008 *Phys. Rev. B* **77** 115117.

IOP Conf. Series: Journal of Physics: Conf. Series **868** (2017) 012004 doi:10.1088/1742-6596/868/1/012004

"Entropics": Science and engineering of caloric phenomena related to itinerant-electron magnetism and spin fluctuations

Asaya Fujita[1*] and Koshi Takenaka[2]

1. Inorganic Functional Materials Research Institute, Department of Materials and Chemistry, National Institute of Advanced Industrial Science and Technology (AIST Chubu), Nagoya 465-8560, Japan
2. Department of Applied Physics, Graduate School of Engineering, Nagoya University, Nagoya 464-8603, Japan

* asaya-fujita@aist.go.jp

Abstract In this paper, the relationship between giant caloric phenomena and itinerant-electron magnetism is examined in order to construct of a newly proposed concept "Entropics", which is a fusion of science and technology with the objective of solving and controlling entropic phenomena. An anomalous hall resistivity is present in the paramagnetic state of the $La(Fe_{0.88}Si_{0.12})_{13}$ magnetocaloric compound. Further, its coefficient exhibits a Curie-Weiss type temperature dependence, indicating the existence of disordered local moment, even though the Rhodes-Wohlfarth (RW) ratio reveals that the magnetic feature in the system is an itinerant-electron type. In addition, the correlation between the magnitude of the transition entropy change of the itinerant-electron metamagnetic transition and the RW ratio was observed. In the Mn_3GaN barocaloric compound, the transition entropy of the first-order antiferromagnetic-paramagnetic phase transition marginally depends on the external pressure, in contrast to the data for $Gd_5Ge_2Si_2$. The origin of this tendency is phase stability against the pressure, as opposed to large volume change at the transition temperature, which results in an enhancement of the barocaloric effect. The influence of topological frustration is also distinguished by comparing it with that of other Mn-based antiperovskite compounds.

1. Introduction

The recent development of thermal management technologies has resulted in many new viewpoints in fundamental and applied solid-state physics, such as spincaloritronics [1] and spin entropy-driven thermoelectricity [2]. Various giant caloric effects owing to latent heat of the solid-state first-order phase transitions are also attracting attention because of the strong prospect of realizing a solid-state heat pump/ refrigeration [3].

With respect to the kind of external field used as a trigger for the phase transition, giant caloric phenomena are classified into categories such as magnetocaloric (MC), baro- (BC) or elast-caloric, and electrocaloric (EC) effects [4-9]. These effects naturally contain common features from a thermodynamic viewpoint and, sometimes, have common physical issues. To treat these issues, we propose the concept of "Entropics"—a fusion of science and engineering developed with the objective of treating such entropic phenomena. In particular, for the strong social demands on the usage of ubiquitous elements, 3d transition metal-based materials are favorable in application and, as a consequence, analyses and control of spin fluctuations in finite-temperature magnetism [10] have become very important in metallic magnetic substances. In this paper, we introduce various examples with striking features in relation between caloric phenomena, such as MC and EC effects, and itinerant-electron magnetism.

2. Results and Discussions

$La(Fe_xSi_{1-x})_{13}$ is a typical itinerant-electron type magnetocaloric material which exhibits an itinerant-electron metamagnetic (IEM) transition above the first-order thermally induced transition at the Curie temperature T_C [7, 11]. The IEM transition appears only at a high Fe concentration range above x = 0.86, where the value of T_C lies at approximately 200 K. Further, hydrogen absorption increases in T_C and the value of T_C can be tuned by hydrogen concentration; e.g., the value varies from 195 K for y = 0 to 332 K for y = 1.3 in $La(Fe_{0.88}Si_{0.12})_{13}H_y$ [7].

Both theoretically and fundamentally, the paramagnetic state as a zero-field state of the IEM transition tends to appear at a maximum in the temperature dependence of susceptibility [12, 13]. This phenomenon is fundamentally explained by spin fluctuations in a (exchange-enhanced) Pauli paramagnetic state [13]. On the other hand, no maximum susceptibility is observed in $La(Fe_xSi_{1-x})_{13}$ except that measurement under the hydrostatic pressures [14], and temperature dependence of paramagnetic susceptibility χ appears to obey the Curie-Weiss (CW) law [14, 15]. In order to verify the thermal behavior of the paramagnetic state, we conducted Hall resistivity ρ_{xy} measurements above T_C. Figure 1 shows the temperature T dependence of the slope of R_H derived from the range close to the zero-magnetic field. As shown in the inset in Fig. 1, the magnetic field dependence of ρ_{xy} is very similar to the magnetization curve, which is featured as a nonlinear increase in magnetization due to the IEM transition. This means that the main part of the observed ρ_{xy} consists of the anomalous Hall effect (AHE) [16]. It is natural for the ferromagnetic state after the IEM transition to induce the AHE, because it contains the spontaneous local moment. Further, the variation of R_H in the paramagnetic state shown in Fig. 1 clearly corresponds to the T^{1} function denoted by the solid line, indicating that a

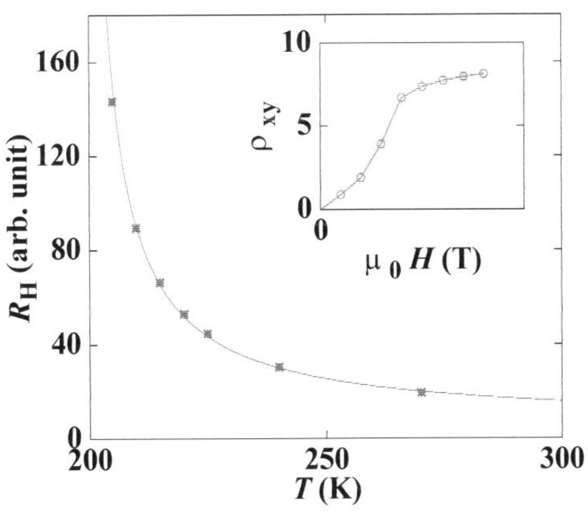

Figure 1. Temperature T dependence of anomalous Hall coefficient R_H in the paramagnetic state for $La(Fe_{0.88}Si_{0.12})_{13}$. Inset shows the magnetic field dependence of the Hall resistivity ρ_{xy} at 205 K.

CW variation also appears in the ρ_H - T curve, as with the χ - T curve. Although the details of the AHE have not yet been fully developed, the spin-orbit interaction (SOI) is the inevitable factor for appearance of the AHE. This result means that the CW behavior in χ originates not only from the pure spin term but also from the unquenched orbital term. An effective moment value of 3.2 μ_B obtained from the $\chi - T$ curve has been reported, corresponding to a Rhodes-Wohlfarth (RW) ratio [17] of approximately 1.2. The RW ratio larger than unity in the $La(Fe_xSi_{1-x})_{13}$ system signifies the itinerant-electron character of magnetism, while its closeness to unity, in addition to the trace of the SOI, prove the local moment character in the PM state. Such a feature is categorized into the so-called "interpolation" region [18], which is located between the weakly itinerant-electron limit in which the dominant magnetic fluctuations are the long wavelength and small frequency types, and the localized moment limit in which dominant fluctuations are characterized by the directionally randomized single-site polarizations. A disordered local moment (DLM) picture has been proposed to treat the paramagnetic local moment in the interpolation region [19, 20]. From first-principle calculations based on the local density approximation (LDA) together with the coherent potential approximation (CPA) [21], the amplitude of the DLM for $La(Fe_{0.88}Si_{0.12})_{13}$ has been evaluated to be approximately 1.9 μ_B. This value is slightly smaller than the magnetic carrier p_c = 2.3 μ_B, which is determined by the relation $p_c(p_c+2) = p_{eff}^2$. The discrepancy is attributable to an additional contribution from the longitudinal spin fluctuations characteristic of itinerant-electron magnetism.

One of the measures used to evaluate the MC performance is isothermal entropy change, ΔS_M. In general, the value of ΔS_M tends to become maximum at just above T_C and subsequently, a plateau is formed at higher temperature ranges. Because the first-order transition occurs at the point where low-temperature and high-temperature phases balance their free energies, F, $F_{high} - F_{low} = 0$. Then, the relation $S_{high} - S_{low} = \Delta S = U_{high} - U_{low}$ is derived, where U is the internal energy, because $F = U - TS$. This means that the phase transition occurs in order to gain internal energy and the internal energy gain is compensated by releasing the entropy difference of two phases. Among them, the internal energy in the FM state is determined a priori when constituent elements and crystal structure are fixed. Further, the contribution of thermal fluctuations to S in the FM state is relatively smaller than that in the PM (DLM) state. Consequently, the nature of spin fluctuations in the DLM state is one of the key factors governing the magnitude of ΔS_M. Figure 2 shows the maximum value of the isothermal entropy change ΔS_M^{max} plotted against the RW ratio for $La(Fe_{0.88}Si_{0.12-y}Al_y)_{13}$, together with the result of $La(Fe_{0.89}Si_{0.12})_{13}$. An increase in the RW ratio was observed after partial substitution of Al, indicating that the nature of the spin fluctuations shifts towards a long-wavelength type. Meanwhile, the value of ΔS_M^{max} gradually decreases with increasing RW ratio. The results shown in Fig. 2 do not instantly signify a causal relationship between these

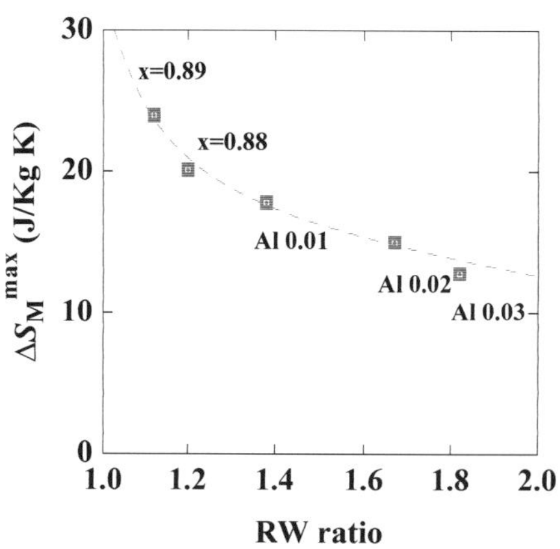

Figure 2. Maximum value of isothermal entropy change ΔS_M plotted against the Rhodes-Wohlfarth ratio for $La(Fe_xSi_{1-x})_{13}$ (x = 0.88 and 0.89) and $La(Fe_{0.88}Si_{0.12-y}Al_y)_{13}$ (y = 0.01 0.02 and 0.03)

two quantities, as the CW behavior means that some kind of interaction between DLMs also influences the fluctuating motions of the DLMs. Note that similar studies into the relationship between transition entropy and the RW ratio for various ferroelectric systems have resulted in data with a tendency very similar to that of the present results [22]. The PM state in the localized moment state has a larger entropy in the magnitude of R ln(2J +1), where R is the gas constant and J is the total angular moment of the localized moment. On the other hand, in general, the energy difference between the FM and PM states in the itinerant-electron system is large [23], resulting in an enlarged difference in U between the FM and PM states. Accordingly, the DLM character in the itinerant-electron magnetism has a certain merit for obtaining a large magnetocaloric effect. Further precise observations for the DLM may result in further advantages in usage of the paramagnetic state of the itinerant-electron magnetism.

The antiferromagnetic Mn_3GaN compound is another example of an itinerant-electron caloric material. The ground state of this compound is the antiferromagnetic (AF) phase with Γ^{5g} spin structure [24], owing to a geometric frustration [24, 25]. At the Neel temperature T_N of about 290 K, the first-order phase transition appears and the paramagnetic (PM) state appears above T_N [26]. A notable feature is volume shrinkage of about 1% accompanied by the AF-PM first-order transition at T_N [26, 27]. Attempts have been made to utilize this large volume change as new negative thermal-expansion materials, which alternate with conventional Invar alloys [28]. Recently, a giant entropy change ΔS was observed when an external pressure p was applied and the value of ΔS became approximately 22 J/kg K under an external pressure of 0.14 MPa [8]. Because a magnetic field-induced caloric change, i.e., the magnetocaloric effect, is not expected in antiferromagnetic systems, the barocaloric effect is a suitable way to assess the phase transition entropy change. From the differential scanning calorimeter (DSC) measurement, the heat absorption (latent heat) across T_N was evaluated at approximately 22 J/kg K; thus, it is apparent that the magnitude of the latent heat governs that of the BC effect.

Figure 3 shows the pressure dependence of the latent heat L for Mn_3GaN, together with that for $Gd_5Ge_2Si_2$. For both materials, the technique for caloric measurements under external pressure is the same as that in our previous paper [8]. $Gd_5Ge_2Si_2$ is known as a typical giant MC material [29] and its magnetic behavior is characterized by Gd 4f localized moment. Its first-order phase transition is fundamentally driven by structural transition, and the FM-PM magnetic transition is induced as a subsidiary. In other words, the magnetic characteristics of these two compounds are completely different. As shown in Fig. 3, the amount of latent heat for the first-order $Gd_5Ge_2Si_2$ apparently decreases with increasing external pressure. A similar decline of the BC effect with

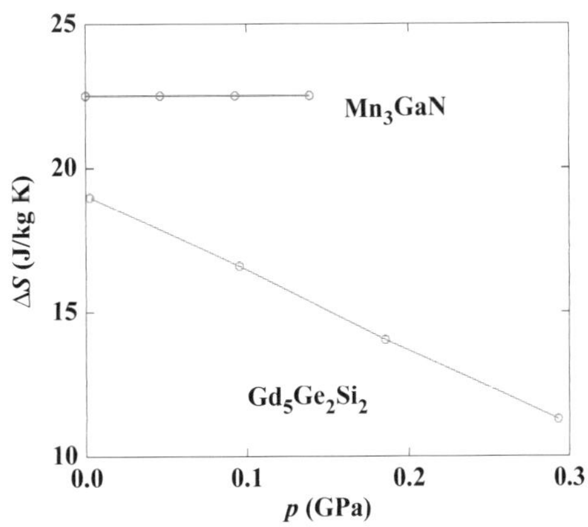

Figure 3. Pressure dependence of the transition entropy ΔS for Mn_3GaN and $Gd_5Ge_2Si_2$

increasing pressure has been observed, and the result has been attributed to a decrease in phonon entropy involved in the total transition entropy [30]. Meanwhile, the value of L scarcely depends on the pressure in these pressure ranges for Mn_3GaN. Such a feature is plausibly explained by the pressure stability of the magnetic state in Mn_3GaN. The appearance of the giant BC effect owing to the pressure stability, or smaller pressure effect on T_N, in contrast to the large volume change at T_N has also been pointed out in this compound [8].

The large volume shrinkage at T_N is caused by a reduction in amplitude due to a change from the ordered local moment (OLM) to the DLM. Meanwhile, the relatively smaller pressure effect of T_N is derived from the small pressure shift of the free-energy balance point. The 3d band contribution to the pressure effect of T_N is relatively smaller than the contribution of spin fluctuations [31]. Accordingly, the difference between the pressure effect on the OLM and DLM is one of the measures used for the pressure effect on the first-order transition temperature. A contrastive example to the present compounds is Mn_3SnC, which exhibits first-order transition from a non-collinear ferrimagnetic to the PM state [26, 32]. In the pressure effect on this transition temperature, T_t is approximately -16 K/GPa [32], whereas the volume change at T_t is approximately 0.2 % [26]. This means that the difference between the OLM and the DLM is not very large, resulting in both the small pressure effect and volume change. The transition entropy change estimated from an extrapolation of ΔS_M to zero-magnetic field becomes a relatively small value of approximately -5 J/kg K [33]. On the other hand, the shrinkage in amplitude from the OLM to DLM is attributable to the influence of topological frustration in Mn_3GaN [8]. In other words, the origin of the shrinkage in the local moment amplitude is different from the band energy gain [23], which is directly correlated with a change in volume, or in Mn-Mn atomic distances. A precise determination of both the OLM and DLM amplitudes and their pressure dependence is inevitable for further detailed discussion. From the viewpoint of engineering, usage of frustration in the AF materials is a notable advantage in the BC effect, and such a case can be a rare example of a positive application of the AF materials to practical applications.

3. Conclusion

In this paper, two examples were presented that helped to clarify the relationship between giant caloric phenomena and itinerant-electron magnetism. Compared to localized moment systems, the controllability of magnetic entropy is more significant in itinerant-electron systems, because of their various characteristics, such as amplitude of local moment, magnetovolume effects, and spin fluctuations. Accordingly, itinerant-electron magnets are very important materials for the concept of "Entropics," which is a fusion of science and technology with the objective of solving and controlling giant entropic phenomena in terms of physics and applications. However, the details of its physics, such as finite-temperature magnetism and the spin-fluctuation model in the "interpolation" region, are still debatable. It is expected that significant progress will be made in these issues in the future.

Acknowledgement
We would like to extend our sincere appreciation to Prof. Yoshinori Takahashi at University of Hyogo, for enlightening us about theoretical visions of spin fluctuations, and for imparting essential advice.

References

[1] Bauer G E W, Saitoh E and van Wees B J Nature Mater. **11** 391 (2012)
[2] Terasaki I Physica B 328 63 (2003)
[3] Crossley S, Mathur N D and Moya X AIP Adv. **5** 067153 (2015)
[4] Moya X, Kar-Narayan S and Mathur N D Nat. Mater. 13, 439 (2014)
[5] Fähler S, Rößler U K, Kastner O, Eckert J, Eggeler G, Emmerich H, Entel P, Müller S, Quandt E and Albe K Adv. Eng. Mater. **14** 10 (2012)
[6] Mañosa L, Planes A and Acet M J. Mater. Chem. A **1** 4925 (2013)
[7] Fujita A, Fujieda S, Hasegawa Y and Fukamichi K, Phys. Rev. B **67** 104416 (2003)
[8] Matsunami D, Fujita A, Takenaka K and Kano M Nature Mater. **14** 73 (2014)

IOP Conf. Series: Journal of Physics: Conf. Series **868** (2017) 012004 doi:10.1088/1742-6596/868/1/012004

[9] Matsunami D and Fujita A Appl. Phys. Lett. **106**, 042901 (2015)
[10] Moriya T *Spin Fluctuations in Itinerant Electron Magnetism* (Springer-Verlag, Berlin) *(1985)*
[11] Fujita A, Akamatsu Y and Fukamichi K J. Appl. Phys. **85** 4756 (1999)
[12] Yamada H Phys. Rev. B 47 11211 (1993)
[13] Ishiyama K, Shinogi A, and Endo K J. Phys. Soc. Jpn. **53** 2456 (1984)
[14] Fujita A, Fukamichi K, Yamada M and Goto T Phys. Rev. B **73** 104420 (2006)
[15] Fujita A, Kano M and Yako H, J. Appl. Phys. **113** 17A924 (2013)
[16] Onoda S, Sugimoto N and Nagaosa N Phys. Rev. Lett. 97, 126602 (2006)
[17] Rhodes R and Wohlfarth E P Proc. Roy. Soc. London A**273** 247 (1963)
[18] Moriya T and Takahashi Y J. Phys. Soc. Jpn. **45** 397 (1978)
[19] Akai H and Dederichs P H Phys. Rev. B **47** 8739 (1993)
[20] Pindor A J, Staunton J, Stocks G M and Winter H J. Phys. F : Metal Phys. **13** 979 (1993)
[21] Fujita A and Yako H Scr. Mater. **67** 578 (2012)
[22] Onodera Y and Sawashima N J. Phys. Soc. Jpn. **60** 1247 (1991)
[23] Stollhoff G, Oles A M and Heine V Phys. Rev. B 41 7028 (1990)
[24] Bertaut E F, Fruchart D, Bouchaud J-P and Fruchart R Solid State Commun. **6** 251 (1968)
[25] Iikubo S, Kodama K, Takenaka K, Takagi H and Shamoto S Phys. Rev. B 77, 020409(R) (2008)
[26] Fruchart D and Bertaut E F J. Phys. Soc. Japan **44** 781 (1978)
[27] Takenaka K, Ichigo M, Hamada T, Ozawa A, Shibayama T, Inagaki T and Asano K Sci. Technol. Adv. Mater. **15** 015009 (2014)
[28] Takenaka K, Asano K, Misawa M and Takagi H Appl.Phys. Lett. **92** 011927 (2008)
[29] Pecharsky V K and Gschneidner Jr. K A Phys. Rev. Lett. **78** 4494 (1997)
[30] Yuce S, Barrio M, Emre B, Stern-Taulats E, Planes A, Tamarit J-L, Mudryk Y, Gschneidner Jr. K A, Pecharsky V K, and Mañosa L Appl. Phys. Lett. **101** 071906 (2012)
[31] Yamada H and Terao K J. Phys. Condens. Matter **6** 10805 (1994)
[32] Kaneko T Kanomata T and Shirakawa K J. Phys. Soc. Jpn. 56 4047 (1987)

Quantum critical behavior in magnetic quasicrystals and approximant crystals

N. K. Sato[1], S. Matsukawa[1], K. Nobe[1], K. Imura[1], K. Deguchi[1], T. Ishimasa[2]

[1]Department of Physics, Nagoya University, Nagoya 464-8602, Japan
[2]Division of Applied Physics, Hokkaido University, Sapporo 060-8628, Japan

E-mail: `kensho@cc.nagoya-u.ac.jp`

Abstract. The electronic states of quasicrystals are believed to be critical, neither extended nor localized. To experimentally establish such the critical state remains a formidable challenge. In the Au-Al-Yb quasicrystal, we observed quantum critical phenomena that are characterized by unconventional critical indices similar to those of Yb-based heavy fermions. In contrast, no divergence was observed in the Au-Al-Yb approximant crystal. These results lead us to suggest that the observed quantum criticality is related to the critical state unique to the quasicrystals. Here we review these results, including the recent observation of the superconductivity in the Tsai-type approximant crystal that is isostructural to the Au-Al-Yb approximant, and argue that the quantum criticality of the quasicrystal results from the combined effect of the quasiperiodicity and the electron correlation.

1. Introduction

Quasicrystals (QCs) and heavy fermions have been one of the main topics in solid state physics and developed independently during the last decades. Recently, a new type of the QC was found in which the thermodynamic properties such as the magnetic susceptibility and the specific heat diverge as $T \to 0$ [1], with the critical indices similar to those of $YbRh_2Si_2$ and $YbAlB_4$ [2, 3]. Study on this novel QC may shed more light on the basic notions of the quantum criticality in the heavy fermion crystals, and an electronic state unique to the QCs as well.

The QC is a metallic alloy that possesses long-range, aperiodic structures (see Fig. 1(a)) with diffraction symmetries such as $m\bar{3}\bar{5}$ forbidden to conventional crystals. Figure 1(b) shows the electron diffraction pattern of our QC sample demonstrating 5-fold symmetry. Note that the Bragg spots are very sharp like in the conventional crystals but not like in amorphous materials, confirming the presence of some kind of the long-range ordering in the QC. Because of this unique structure, we expect the existence of the so-called critical state that is neither extended nor localized. This novel electronic state has long been pursued by experimentalists but has not been established yet. As a promising candidate of the QC critical state, the valence fluctuation QC, $Au_{51}Al_{34}Yb_{15}$, has attracted much attraction; while the QC shows the quantum critical phenomenon, its corresponding periodic crystal called the approximant crystal (AC; see Fig. 1(c)) shows no such divergence and only heavy Fermi liquid behavior at low temperatures [1].

The QC and AC of the present interest consist of atomic decorations (i.e., icosahedral clusters of atoms) called Tsai-type cluster, which is composed of a concentric arrangement of multiple

IOP Conf. Series: Journal of Physics: Conf. Series **868** (2017) 012005 doi:10.1088/1742-6596/868/1/012005

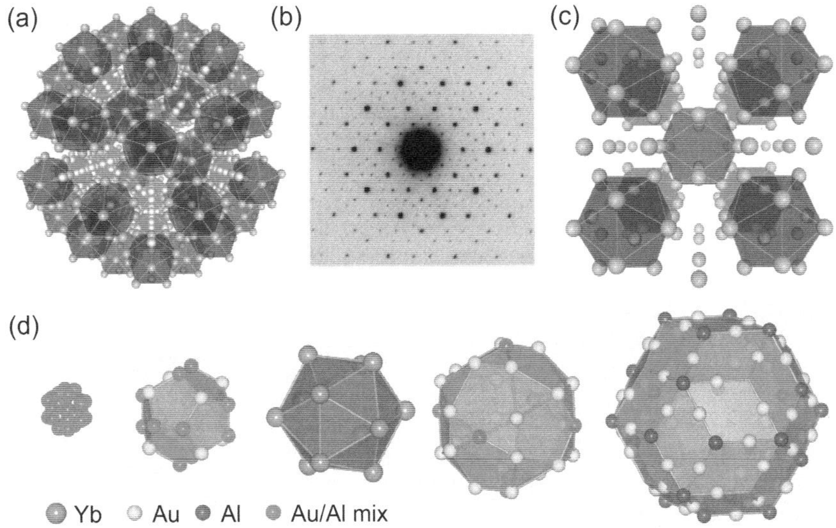

Figure 1. Geometric structure of the Tsai-type quasicrystal (QC) and approximant crystal (AC) [1]. (a) Aperiodic array of icosahedrons in the QC. (b) Diffraction pattern of the QC studied here. (c) Body-centered cubic array of the icosahedrons in the AC. (d) Tsai-type cluster composed of a concentric arrangement of multiple shells.

shells (Fig. 1(d)). At the center of the first shell, there is a polyhedron that consists of 4 atoms in total and points in a various direction. The second shell is a dodecahedron that is composed of mixed Au/Al atoms. The third shell is an icosahedron with Yb ions on the vertex. The fourth and fifth shell are an icosidodecahedron and triacontahedron, respectively. In the recently discovered Au-Ge-Yb superconducting ACs [4], there are two types of the crystal structure; one contains the tetrahedron as described above [5], and the other contains the single Yb atom at the center of the dodecahedron [6]. These Tsai-type clusters are arranged quasiperiodically in the QC (Fig. 1(a)), while in the 1/1 AC, they are arranged periodically to form a body-centered cubic (bcc) structure (space group: $Im\bar{3}$) (Fig. 1(c)).

Before describing the geometrical relationship between the QC and the AC, we present the prescription to construct the quasiperiodicity from the conventional lattice [7]. Figure 2(a) illustrates the square lattice in the plane of X_1 and X_2. Let us separate the two dimension (2D) space into two orthogonal subspaces, $X_\parallel$ ("physical space") and $X_\perp$ ("complementary space"), by rotating the original set of the axes by some angle θ such that $\tan\theta = 1/\tau$ (where $\tau = (1+\sqrt{5})/2$ is the golden ratio). In order to produce atoms in the physical space, we introduce the atomic basis (sometimes called "acceptance region", "superatom", or "hyperatom") as represented by the heavy line segments perpendicular to $X_\parallel$. The intersection of the basis segments with the $X_\parallel$ axis locates the position of atoms in the physical space. We find that the atomic density along $X_\parallel$ is described by a Fibonacci sequence of long (L) and short (S) interatomic spacings, $L, S, L, L, \cdots$, corresponding to the quasiperiodicity in the QC. It may be understood that by extending this prescription, an aperiodic array of atoms in 3Ds can be obtained from a periodic lattice, "hypercrystal", in 6Ds.

Now we show that the AC structure results from introducing a "linear phason strain" in the QC hypercrystal, using a simple example shown in Fig. 2(b). The "phason displacement" w is a displacement of the atomic basis along the $X_\perp$ axis, and the "phason strain" is defined by $dw/dX_\parallel$; then, the linear phason strain is characterized by a constant A such that

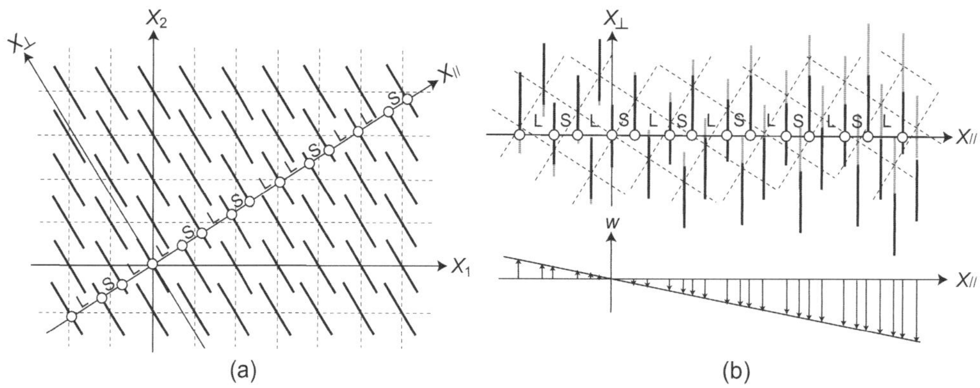

Figure 2. (a) Construction of an aperiodic array of atoms in 1D from a periodic square lattice in 2Ds. X_1 and X_2 axes denote the square lattice. The physical space ($X_\parallel$) represents a 1D cut through the 2D periodic lattice. The sequence of L and S on the $X_\parallel$ axis corresponds to the Fibonacci sequence. (b) Construction of AC from QC. The light-gray symbols in the upper panel denote the atomic basis on the square lattice of the QC. By inducing the linear phason strain w shown in the lower panel, the periodic sequence with a period LS (L and S being the length between the black atomic bases) emerges on the $X_\parallel$ axis.

$w = AX_\parallel$ as shown in the lower panel. The constant A is expressed in terms of the Fibonacci sequence as $A = (F_{n+1} - F_n\tau)/(F_{n+1}\tau + F_n)$, where F_n is a successive Fibonacci number; $F_0 = 0, F_1 = 1, F_2 = 1, F_3 = 2, \cdots$. In the case of Fig. 2(b), A corresponds to $n = 1$, which gives rise to the displacement from the light-gray to black segments shown in the upper panel, and the resulting arrangement of the atomic basis along the $X_\parallel$ becomes periodic, $L, S, L, S, \cdots$. This is just a regular crystal with the unit cell of LS, and called 1/1 AC. In the case of $n = 2$, the interatomic spacings become periodic again, but the unit cell is LSL. This is called 2/1 approximant; the number 2 and 1 corresponds to the number of L and S in the unit cell, respectively. Note that the higher order AC (characterized by the larger n) has the larger unit cell size. As a result, in an ideal case, the QC is regarded geometrically as a limit of the rational AC with the infinitely large unit cell, in other words, the AC approaches the QC as $A \to 0$.

The 3D AC is characterized by the conventional cubic lattice parameter $a_{3\mathrm{D}}$ as usual, whereas the 3D QC is characterized by the 6D lattice parameter $a_{6\mathrm{D}}$. (For our case, the Au-Al-Yb QC and AC have $a_{6D} = 0.7448$ nm and $a_{3\mathrm{D}} = 1.4500$ nm, respectively.) These lattice parameters are connected by the relation, $a_{3\mathrm{D}} = Ba_{6\mathrm{D}}$, where $B = \sqrt{2/(\tau+2)}(F_{n+1}\tau + F_n)$. For the 1/1 AC, B is approximately equal to 1.946. For the higher order ACs, the same relation holds but B takes a different value from the above number. Remembering these in mind, we discuss the physical properties of the QCs and ACs in the next section.

2. Results and Discussion

Figure 3(a) shows the inverse magnetic susceptibility of $Au_{49}Al_{34}Yb_{17}$ QC (referred to as Au-Al-Yb QC hereafter) as a function of temperature [1, 8], which shows a Curie-Weiss form above ~100 K. The linear slope yields the effective moment as $p_{\mathrm{eff}} = 3.91\mu_{\mathrm{B}}$ per Yb ion (where μ_{B} is the Bohr magneton). This value is smaller than that of the free Yb^{3+} ion, 4.54 μ_{B}, indicating that the Yb-ion of the QC is in between Yb^{3+} and Yb^{2+}, in agreement with X-ray absorption near edge structure experiments [9].

As $T \to 0$, the magnetic susceptibility of the QC shows the quantum critical behavior, $\chi \propto T^{-0.51}$, at $H = 0$ as shown in Fig. 3(b). The electronic specific heat coefficient also shows

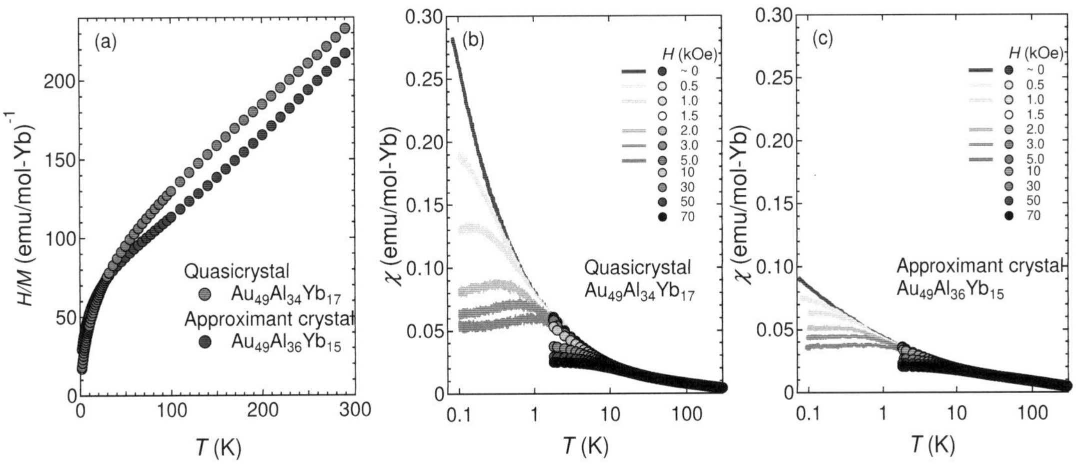

Figure 3. Temperature dependences of the magnetic susceptibility of the Au-Al-Yb QC and AC [1, 8]. (a) The inverse magnetic susceptibility of QC and AC. (b) The dc and ac magnetic susceptibility of QC under magnetic fields indicated in the figure. (c) The dc and ac magnetic susceptibility of AC under magnetic fields indicated in the figure.

the divergent behavior [1], $C/T \propto -\ln T$, where C is specific heat.

One may think that this unconventional low-T behavior might be caused by the random occupation of the Au and Al atoms in the Tsai-type clusters [10]. To examine this possibility, we measured the susceptibility of the 1/1 AC, $Au_{49}Al_{36}Yb_{15}$, whose composition is very similar to that of the QC. Note that it also contains the same type of randomness as the QC [11]. While the AC shows the intermediate valence nature ($p_{eff} = 3.96\mu_B$) at high temperatures similarly to the QC (Fig. 3(a)), it shows a tendency to saturate at low temperatures in contrast to the QC (Fig. 3(c)). (The specific heat also shows the Fermi liquid feature at low temperatures [1], in contrast to the divergent feature in the QC mentioned above.) If the non-Fermi liquid behavior of the QC would be due solely to the randomness, then the similar divergence should be observed in the AC. However, this is not the case. Therefore, it is very unlikely that the unusual low-T behavior of the QC is caused by the randomness.

Instead, the quantum criticality of the QC seems to be reasonably ascribed to the presence of the quasiperiodicity, because it is absent in the AC. However, one may think that in the case of the QC, the quantum critical point would appear accidentally at ambient pressure. To check this possibility, we measured the high-pressure magnetic susceptibility. For the QC, the divergent behavior survives at the highest pressure accessible in the present experiment ($\sim$ 2.8 GPa) (not shown here). In contrast, the AC shows the heavy-fermion like feature: The low-temperature susceptibility increases with pressure, and at a high pressure exceeding 2 GPa, a cusp-like anomaly emerges at a very low temperature below 100 mk (not shown here). It is open whether this is ascribed to the long-range ordering or the spin-glass like transition, but we failed to find in the AC the quantum critical point where the susceptibility diverges as $T \to 0$. The different response to the hydrostatic pressure between the QC and the AC implies that the quantum criticality even at ambient pressure is not fortuitous but inherent to the QC, and confirms that the quasiperiodicity plays a role in the emergence of the quantum criticality of the QC.

The question to be addressed here is if all the QCs would show the quantum criticality. To examine this, we prepared the Tm-based samples isostructural to the Yb-based ones, and measured the magnetic susceptibility [12]. The temperature dependences of the Au-Al-Tm

IOP Conf. Series: Journal of Physics: Conf. Series **868** (2017) 012005 doi:10.1088/1742-6596/868/1/012005

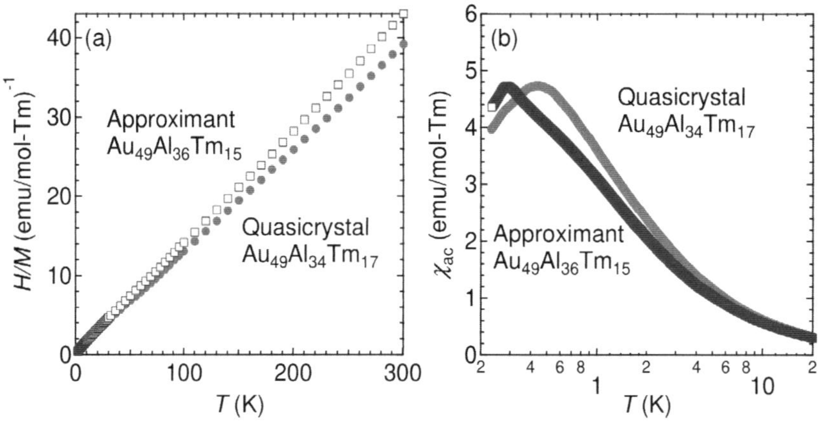

Figure 4. Temperature dependence of magnetic susceptibility of the Au-Al-Tm QC and AC [12]. (a) The inverse dc magnetic susceptibility measured at $H = 500$ Oe. (b) Ac magnetic susceptibility measured at zero dc field at low temperatures. Note that the broad peak emerges at $T_f \simeq 0.43$ and 0.28 K for the QC and the AC, respectively.

QC and AC are shown in Fig. 4(a). The $1/\chi(T)$ curve shows the linear behavior over a wide temperature range, and the slope above ~150 K yields $p_{\rm eff} = 7.8$ and 7.4 μ_B/Tm for the QC and the AC, respectively. These $p_{\rm eff}$ values are close to the free ion value of Tm^{3+} (7.54 μ_B), indicating that their $4f$ electrons are localized in the real space. At low temperatures, both the QC and AC show a transition into a spin-glass like state as suggested by a broad peak at a temperature T_f below 1 K (Fig. 4(b)). Note that the specific heat also shows a broad peak at a temperature close to T_f [12]. It should be mentioned that the Curie-Weiss behavior continues to low temperatures down to T_f. From these results, we conclude that the valence fluctuation is also needed for the emergence of the quantum criticality in the Yb-based QC.

The importance of the valence fluctuation as a result of the hybridization effect may be confirmed from the substitution effect of the constituent elements on the physical properties. We prepared the $Au_{44}Ga_{41}Yb_{15}$ and $Ag_{47}Ga_{38}Yb_{15}$ AC, which were found to belong to the 1/1 AC with the lattice constant of $a_{3D} = 1.4527$ and 1.4687 nm, respectively. As may be expected from the fact that these lattice constants are larger than that of the Au-Al-Yb AC, 1.4500 nm, the two substituted ACs possess no localized magnetic moment and hence their Yb valence is divalent. From this observation, we suggest that the Au-Al-Yb QC and AC are located near the border of the divalent and trivalent states of the Yb ion [8].

We also synthesized $Au_{64.0}Ge_{22.0}Yb_{14.0}$ and $Au_{63.5}Ge_{20.5}Yb_{16.0}$, which are referred to as Au-Ge-Yb(I) and Au-Ge-Yb(II), hereafter. Our structure analysis showed that the Au-Ge-Yb(I) is a Tsai-type 1/1 AC with $a_{3D} = 1.4724$ nm [4]. For Au-Ge-Yb(II), we were unable to give a structure model due to the lack of a single phase sample, but it was reported by Gebresenbut *et al.* that Au-Ge-Yb(II) is also a Tsai-type 1/1 AC [6]. These two ACs have a slight difference in the crystal structure as mentioned in Introduction: The cluster center of the Au-Ge-Yb AC(I) is a tetrahedron, while in the Au-Ge-Yb AC(II), a single Yb ion occupies the cluster center; as a result, 12 and 13 Yb ions are contained in the Tsai-type cluster in the Au-Ge-Yb AC(I) and AC(II), respectively. We note that the Au-Ge-Yb AC(I) and AC(II) show superconductivity below 0.68 and 0.36 K, respectively. From a detailed analysis of the magnetism, we found that while the Yb ions on the vertex of the icosahedron are divalent, the cluster-center Yb ion is likely of intermediate valence. These results suggest that the hybridization effect plays an important role in the physical properties of the materials investigated here. The interplay between the

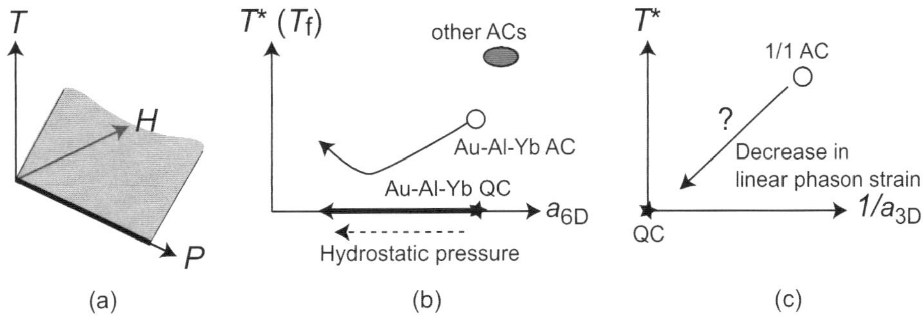

Figure 5. (a) Schematic illustration of T-P-H phase diagram of the Au-Al-Yb QC. The hatched plane marks T^* below which the Fermi liquid is observed. (b) Schematic illustration of the characteristic temperature T^* (or the transition temperature T_f into a short-range or long-range ordered state) as a function of the 6D lattice parameter a_{6D}. (c) T^* as a function of the inverse of the 3D lattice parameter a_{3D}. Note that the QC can be regarded as the AC with $a_{3D} \to \infty$.

magnetism and the superconductivity remains to be resolved in the future.

Let us study the magnetic field effect on the Au-Al-Yb QC. As was shown in Figs. 3(b) and (c), the application of uniform magnetic field H suppresses the growing of the magnetic susceptibility with lowering temperature in both the QC and the AC, resulting in a field-induced Fermi liquid state below a characteristic temperature T^* with an enhanced Pauli susceptibility. According to our preliminary measurements, the strong sensitivity to the magnetic field was also observed at high pressures. Such the magnetic field effect is often observed in the heavy fermions, and the AC is considered as a heavy fermion system. Therefore, the nonmagnetic ground state such as the Kondo state is likely formed in the QC.

Finally we summarize the characteristics of the Au-Al-Yb QC comparing the ACs. In the 3D space of the temperature, pressure, and magnetic field, the QC is located on the "quantum critical line" at $T = 0$ (see the heavy line on the P axis in Fig. 5(a)). (Note that the quantum critical line running on the zero temperature plane in the phase diagram might be ubiquitous as was theoretically discussed to study a novel high-pressure phase of the itinerant electron ferromagnet $ZrZn_2$ [13]; it may be interesting to compare the experimentally obtained phase diagram of $ZrZn_2$ [14] with the present result.) This results in the plot of T^* vs. a_{6D} (Fig. 5(b)) in which the QC takes a route as indicated by the thick arrow: Here, the application of the hydrostatic pressure decreases a_{6D} like the conventional lattice parameter a_{3D}. Remembering that $a_{6D} \propto a_{3D}$, the Au-Al-Yb AC moves as indicated by the thin arrow. Then, we find that the traces of the QC and the AC do not cross each other. Other ACs than the Au-Al-Yb AC occupy a position of high T^* values, reflecting that they behave like a conventional metal. This viewgraph may help us to distinguish the Au-Al-Yb QC with the Yb-based ACs. We consider that this difference comes from the difference in the periodicity/quasiperiodicity and the stable/unstable valence, which allows us to conclude that the unconventional quantum criticality of the Au-Al-Yb QC is caused by the combined effect of the quasiperiodicity and the strong electron correlation.

We speculate the material dependence of the quantum critical behavior as shown in Fig. 5(c), in which the QC is located at a limit of zero linear phason strain. To confirm this speculation would give a direct evidence for the above conclusion. The search for the higher-oder ACs like the 2/1 AC is a challenging work in the future.

Acknowledgments
This work was partially supported by Grants-in-Aid for Scientific Research from JSPS, KAKENHI (Nos. 25610094, 26610100, 15H02111, and 1503685).

References
[1] Deguchi K, Matsukawa S, Sato N K, Hattori T, Ishida K, Takakura H and Ishimasa T 2012 *Nat. Mater.* **11** 1013
[2] Gegenwart P, Westerkamp T, Krellner C, Tokiwa Y, Paschen S, Geibel C, Steglich F, Abrahams E and Si Q 2007 *Science* **16** 969
[3] Matsumoto Y, Nakatsuji S, Kuga K, Karaki Y, Horie N, Shimura Y, Sakakibara T, Nevidomskyy A H and Coleman P 2011 *Science* **331** 316
[4] Deguchi K, Nakayama M, Matsukawa S, Imura K, Tanaka K, Ishimasa T and Sato N K 2015 *J. Phys. Soc. Jpn.* **84** 023705
[5] Deguchi K, Nakayama M, Matsukawa S, Imura K, Tanaka K, Ishimasa T and Sato N K 2015 *J. Phys. Soc. Jpn.* **84** 015002
[6] Gebresenbut G H , Tamura R, Eklof D and Gomez C P 2013 *J. Phys.: Condens. Matter* **25** 135402
[7] Goldman A I and Kelton K F 1993 *Rev. Mod. Phys.* **65** 213
[8] Matsukawa S, Tanaka K, Nakayama M, Deguchi K, Imura K, Takakura H, Kashimoto S, Ishimasa T and Sato N K 2014 *J. Phys. Soc. Jpn.* **83** 034705
[9] Watanuki T, Kashimoto S, Kawana D, Yamazaki T, Machida A, Tanaka Y and Sato T J 2012 *Phys. Rev.* B **86** 094201
[10] Löhneysen H, Rosch A, Vöjta M and Wolfle P 2007 *Rev. Mod. Phys.* **79** 1015
[11] Ishimasa T, Tanaka Y and Kashimoto S 2011 *Philos. Mag.* **91** 4218
[12] Nakayama M, Tanaka K, Matsukawa S, Deguchi K, Imura K, Ishimasa T and Sato N K 2015 *J. Phys. Soc. Jpn.* **84** 024721
[13] Yamaji Y, Misawa T and Imada M 2007 *J. Phys. Soc. Jpn.* **76** 063702
[14] Kabeya N, Maekawa H, Deguchi K, Kimura N, Aoki H and Sato N K 2012 *J. Phys. Soc. Jpn.* **81** 073706

IOP Conf. Series: Journal of Physics: Conf. Series **868** (2017) 012006 doi:10.1088/1742-6596/868/1/012006

Quantum Spin Liquid in the Kagome-Lattice Antiferromagnet and Related Systems

Hiroki Nakano[1] **and Tôru Sakai**[1,2]

[1]Graduate School of Material Science, University of Hyogo, Hyogo 678-1297, Japan
[2]Japan Atomic Energy Agency, SPring-8, Hyogo 679-5148, Japan

E-mail: sakai@spring8.or.jp

Abstract. The S=1/2 kagome-lattice antiferromagnet is investigated using the numerical diagonalization up to the 42-spin cluster. The analysis of the field derivative of the magnetization at the zero magnetization indicates that the magnetic excitation ot the system is gapless. It is consistent with our previous finite-scaling analysis of the spin gap.

1. Introduction

The $S = 1/2$ kagome-lattice antiferromagnet is one of interesting frustrated quantum spin systems. The system is supposed to exhibit the quantum spin liquid in the ground state[1, 2, 3, 4, 5, 6, 7, 8, 9, 10, 11, 12], which was proposed as an origin of the high-temperature superconductivity[13]. Experimental studies to observe a novel spin liquid phase have been accelerated, since discoveries of several candidate materials; the herbertsmithite[14, 15], the volborthite[16, 17] and the vesignieite[18] for the kagome lattice. The spin gap is an important physical quantity to characterize the spin liquid behavior. Whether the $S = 1/2$ kagome-lattice antiferromagnet is gapless or has a finite spin gap, is still an unsolved issue, because any recently developed numerical calculation methods are not enough to determine it in the thermodynamic limit. The quantum Monte Carlo (QMC) simulation has a so-called negative sign problem due to the frustration. The density matrix renormalization group (DMRG) method is not good because of the two-dimensionality of the system. Our large-scale numerical diagonalization up to the 42-spin cluster and a finite-size scaling analysis indicated that the $S = 1/2$ kagome-lattice antiferromagnet is gapless in the thermodynamic limit[19]. It is consistent with the U(1) Dirac spin liquid theory of the kagome-lattice antiferromagnet[8]. It was also supported by a variational method[20, 21]. On the other hand, some DMRG calculations[22, 23, 24] indicated a finite spin gap in the thermodynamic limit. Although the estimated magnitude of the spin gap is dependent on the method, they are qualitatively consistent with the gapped Z_2 topological spin liquid theory[1]. It was also supported by a numerical diagonalization study[25]. Although the recent neutron scattering experiment of the single crystal of the herbertsmithite[26, 27] suggested that the system is gapless, the spin gap issue of the kagome-lattice antiferromagnet is still theoretically controversial.

In this paper, we analyze the field derivative of the magnetization at the zero magnetization, using the numerical diagonalization up to the 42-spin cluster, to determine whether the kagome-lattice antiferromagnet is gapless or gapped.

IOP Conf. Series: Journal of Physics: Conf. Series **868** (2017) 012006 doi:10.1088/1742-6596/868/1/012006

2. Model and numerical exact diagonalization

Now we examine the spin gap issue of the $S = 1/2$ kagome lattice antiferromagnet. The Hamiltonian is given by

$$\mathcal{H} = \sum_{\langle i,j \rangle} \mathbf{S}_i \cdot \mathbf{S}_j, \tag{1}$$

where $\langle i,j \rangle$ means all the nearest neighbor pairs on the kagome lattice. For an N-site system, the lowest energy of $\mathcal{H}$ in the subspace where $\sum_j S_j^z = M$ is denoted by $E(N, M)$. Using the numerical exact diagonalization, we have calculated all the values of $E(N, M)$ available for the clusters up to N =42 under the periodic boundary condition. The largest dimension of the $N = 42$ system is 538,257,874,440. To treat such huge matrices in computers, we have carried out parallel calculations using the MPI-parallelized code, which was originally developed in the study of the Haldane-gap issue[28]. Note especially that calculations for the N =42 cluster require the use of K computer, Kobe, Japan.

3. Field derivative of the magnetization

The effect of the applied external magnetic field h is discribed by the Zeeman energy term

$$\mathcal{H}_Z = -h \sum_j S_j^z \tag{2}$$

The energy of $\mathcal{H}$ per site in the thermodynamic limit is defined as

$$\frac{E(N,M)}{N} \sim \epsilon(m) \qquad (N \to \infty) \tag{3}$$

where $m = M/(SN)$ is the magnetization normalized by the saturated magnetization SN. If we assume $\epsilon(m)$ is an analytic function of m, the spin excitation energy would become

$$E(N, M+1) - E(N, M) \sim \frac{1}{S}\left(\epsilon'(m) + \frac{1}{2}\epsilon''(m)\frac{1}{NS} + \cdots\right) \tag{4}$$

Thus, this equation gives the quantity corresponding to the width of the magnetization plateau at m as follows,

$$(E(N, M+1) - E(N, M)) - (E(N, M) - E(N, M-1)) \sim \epsilon''(m)\frac{1}{NS^2} \tag{5}$$

Minimizing the energy of the total Hamiltonian $\mathcal{H} + \mathcal{H}_Z$, the ground state magnetization curve is derived by

$$h = \epsilon'(m)/S \tag{6}$$

The field derivative of the magnetization is defined as

$$\chi \equiv \frac{dm}{dh} = \frac{S}{\epsilon''(m)} \tag{7}$$

If we assume $\chi \neq 0$, namely $\epsilon''(m)$ is finite, the magnetization plateau at m would vanish in the thermodynamic limit, because of (5). Thus a necessary condition for the existence of a magnetization plateau at m is $\chi = 0$ in the thermodynamic limit. Now we apply this argument for the spin gap. We should examine the case of $m = 0$. In this case, the equation (5) can be rewritten as

$$2\Delta_N \sim \epsilon''(0)\frac{1}{NS^2} \tag{8}$$

where $\Delta_N = E(N, 1) - E(N, 0)$ is the spin gap for an N-spin cluster. Thus a necessary condition of the finite spin gap would be $\chi = 0$ at $m = 0$ in the thermodynamic limit.

In the next section, we confirm the validity of this condition for typical gapless and gapped spin systems.

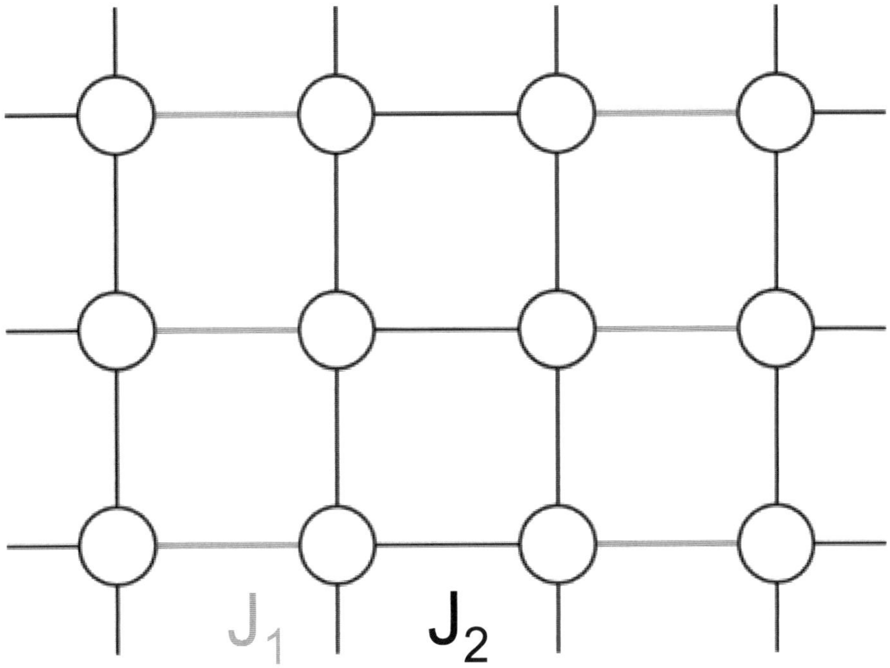

Figure 1. Square-lattice antiferromagnet with the dimerization. The green and black lines denote bonds of interactions J_1 and J_2, respectively. The case of $J_2/J_1 = 1$ corresponds to the square-lattice antiferromagnet without the dimerization.

4. Square-lattice Heisenberg antiferromagnet with and without dimerization

As typical gapless and gapped spin systems, we consider the $S = 1/2$ square-lattice Heisenberg antiferromagnet, and the one with the dimerization, respectively. These models are shown in Figure 1. The ratio of the two antiferromagnetic exchange interactions is defined as $\alpha = J_2/J_1$. The previous QMC simulation study[29] revealed that the critical ratio is $\alpha_c = 0.52337(3)$ and the system is gapless for $\alpha > \alpha_c$, while gapped for $\alpha < \alpha_c$. Using the numerical diagonalization, we investigate the two typical cases; $J_2/J_1 = 1$ (gapless) and $J_2/J_1 = 0.2$ (gapped). The ground state magnetization curves for $J_2/J_1 = 1$ and $J_2/J_1 = 0.2$ are shown in Figures 2 and 3, respectively. The former seems gapless, while the latter clearly looks gapped at $m = 0$. The field derivatives of the magnetization χ for $J_2/J_1 = 1$ and $J_2/J_1 = 0.2$ are shown in Figures 4 and 5, respectively. In both cases the system size dependence of χ at $m = 0$ looks small. We can easily expect that χ is finite for $J_2/J_1 = 1$, while $\chi = 0$ for $J_2/J_1 = 0.2$ in the thermodynamic limit. χ at $m = 0$ plotted versus $1/N$ in Figure 6 for $J_2/J_1 = 1$ and Figure 7 for $J_2/J_1 = 0.2$ clearly confirms these features. The result justifies the argument in the previous section that χ is finite for the gapless system, while $\chi = 0$ for the gapped one in the thermodynamic limit.

5. Kagome lattice antiferromagnet

In this section, we investigate the field derivative of the magnetization χ for the $S = 1/2$ kagome-lattice antiferromagnet. The ground state magnetization curves for N=36, 39 and 42 are shown in Figure 8; the results for N=39 and 42 were originally presented in [30] and [31], respectively. From the magnetization curves, it is difficult to determine whether the system is gapless or gapped. Next we show the magnetization dependence of the calculated χ for N=42, 39 and 36 in Figure 9. The behaviors around $m = 0$ are magnified in Figure 10. At least for N=36 and 42,

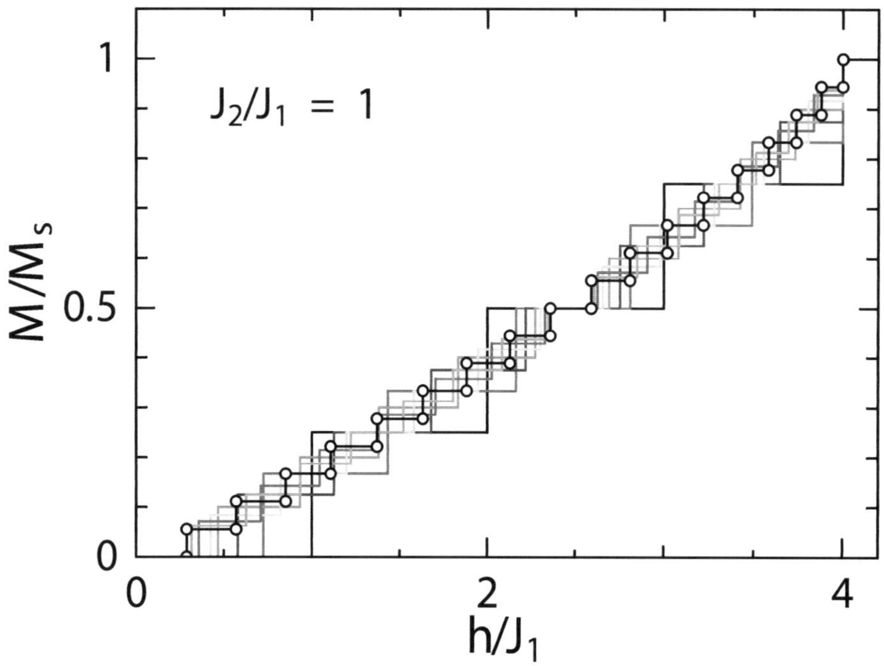

Figure 2. Magnetization curves of the square-lattice antiferromagnet for several system sizes up to $N = 36$. The results for $N = 8$, 12, 16, 20, 24, 28, and 32 are denoted by black, red, dark-blue, green, yellow, violet, light-blue lines without symbols, respectively; the result for $N = 36$ is denoted by black lines with circles.

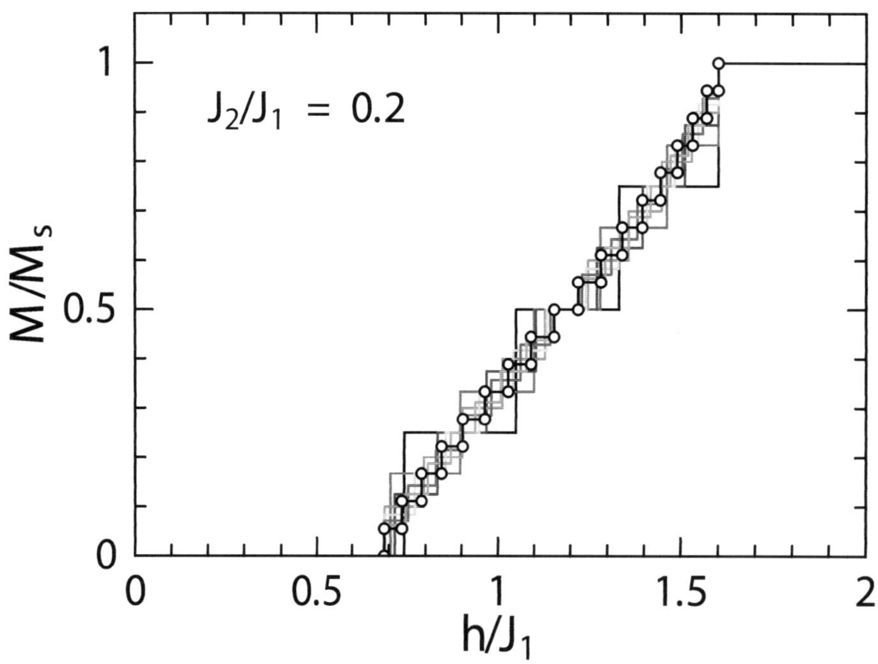

Figure 3. Magnetization curves of the square-lattice antiferromagnet with the dimerization of $J_2/J_1 = 0.2$ for several system sizes up to $N = 36$. Lines and symbles are the same as in Fig. 2.

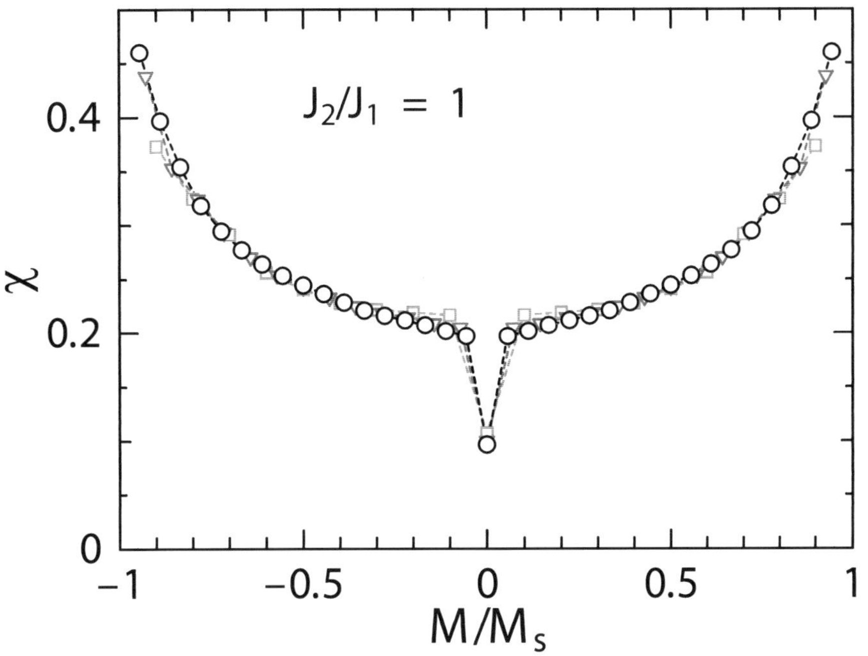

Figure 4. Magnetization dependence of χ of the square-lattice antiferromagnet for several system sizes. The results for N =36, 28, 20 are denoted by black circles, violet inversed triangles, and green squares, respectively.

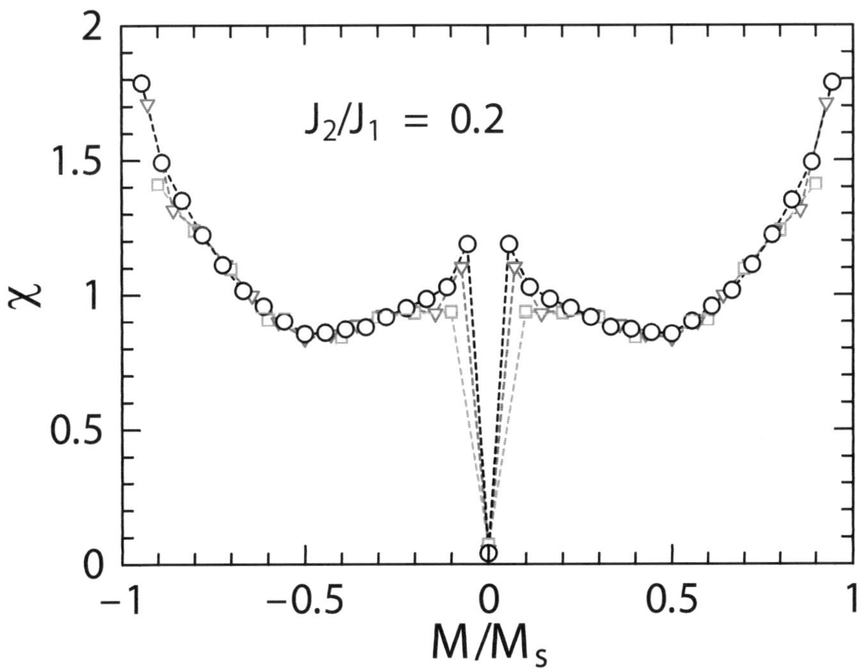

Figure 5. Magnetization dependence of χ of the square-lattice antiferromagnet with the dimerization of $J_2/J_1 = 0.2$ for several system sizes. Lines and symbles are the same as in Fig. 4.

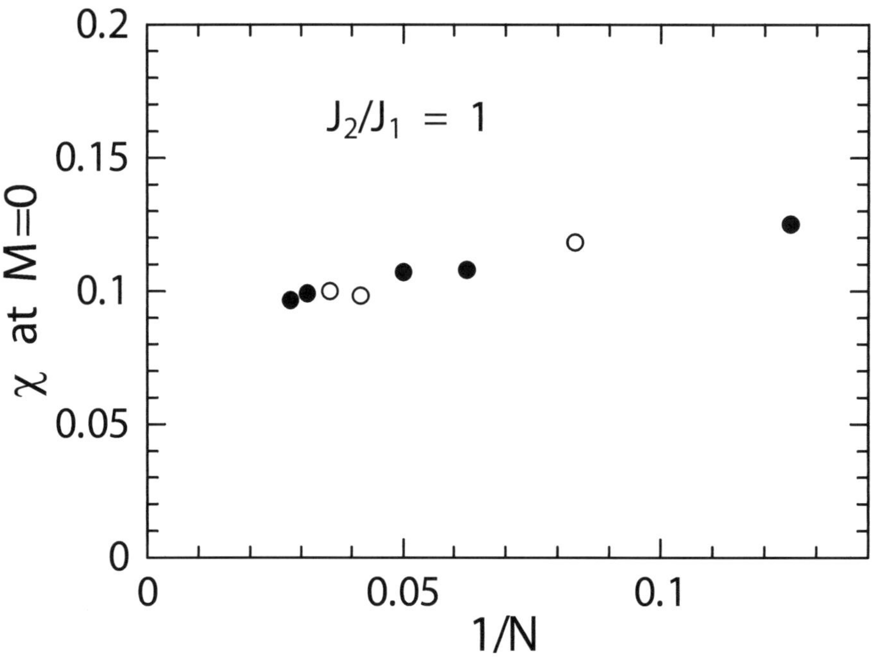

Figure 6. χ plotted versus $1/N$ for the square-lattice antiferromagnet ($J_2/J_1 = 1$). The results are presented for the cases up to $N = 36$. Closed symbols denote the results for the cases when the shape of the finite-size clusters are squares. When a finite-size cluster for a given N cannot form a square even if the square is tilted, on the other hand, the cluster forms a parallelogram; the results are given by open symbols. The extrapolated value in the thermodynamic limit seems to be finite ($\chi \neq 0$).

the size dependence of χ at $m = 0$ is very small. χ at $m = 0$ is plotted versus $1/N$ for N=42, 36, 30, 24, 18, 12 in Figure 11. Although the system size dependence exhibits a slight oscillation, it clearly indicates that χ at $m = 0$ is still finite in the thermodynamic limit. Thus the system does not meet the condition for the finite spin gap. It should be one of strong evidences to justify that the $S = 1/2$ kagome-lattice antiferromagnet is gapless.

6. Conclusion

The spin gap issue of the kagome-lattice antiferromagnet is investigated using the numerical diagonalization up to $N = 42$. The analysis of the field derivative of the magnetization χ gives one of strong evidences of the gapless feature.

Acknowledgments

This work has been partly supported by Grants-in-Aids for Scientific Research (B) (No.23340109) and (C) (Nos. 24540348 and 26400387) from the Ministry of Education, Culture, Sports, Science and Technology of Japan (MEXT). This research used computational resources of the K computer provided by the RIKEN Advanced Institute for Computational Science through the HPCI System Research projects (Project IDs: hp130070, hp130098 and hp150024). We further thank the Supercomputer Center, Institute for Solid State Physics, University of Tokyo; the Cyberscience Center, Tohoku University; and the Computer Room, Yukawa Institute for Theoretical Physics, Kyoto University; the Department of Simulation Science, National Institute for Fusion Science; Center for Computational Materials Science, Institute for Materials Research,

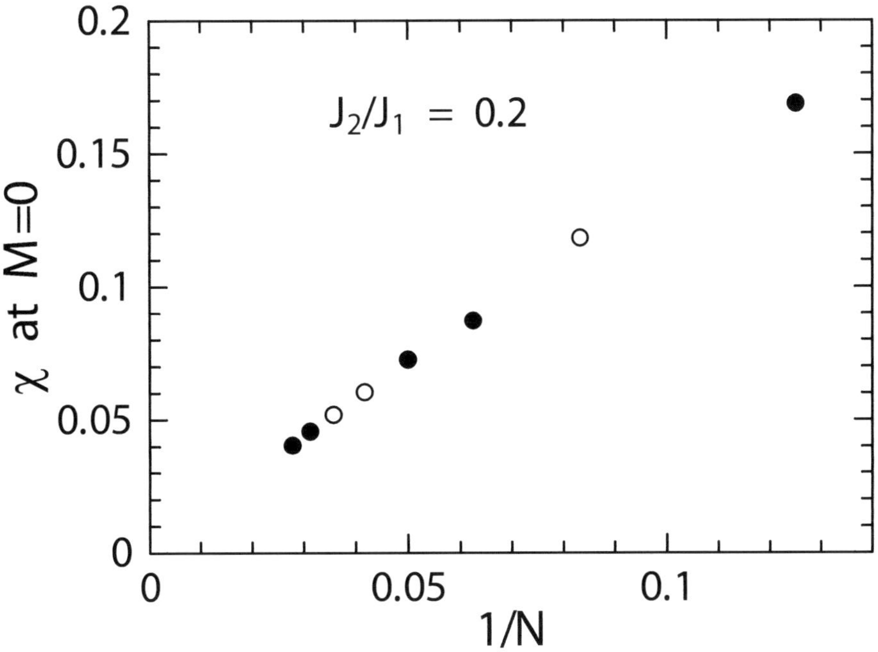

Figure 7. χ plotted versus $1/N$ for the dimerized square-lattice antiferromagnet ($J_2/J_1 = 0.2$). The extrapolated value in the thermodynamic limit seems to be zero. Symbols are the same as in Fig. 6.

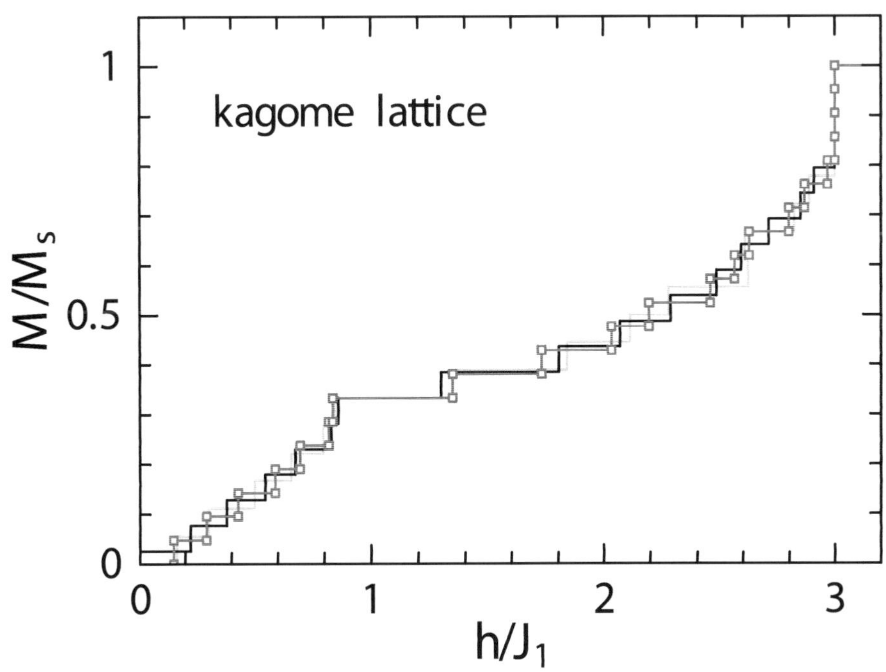

Figure 8. Magnetization curves of the kagome-lattice antiferromagnet for N =36, 39 and 42. The results for N =36 and 39 are denoted by green and black lines, respectively; that for N =42 is denoted by red lines with squares.

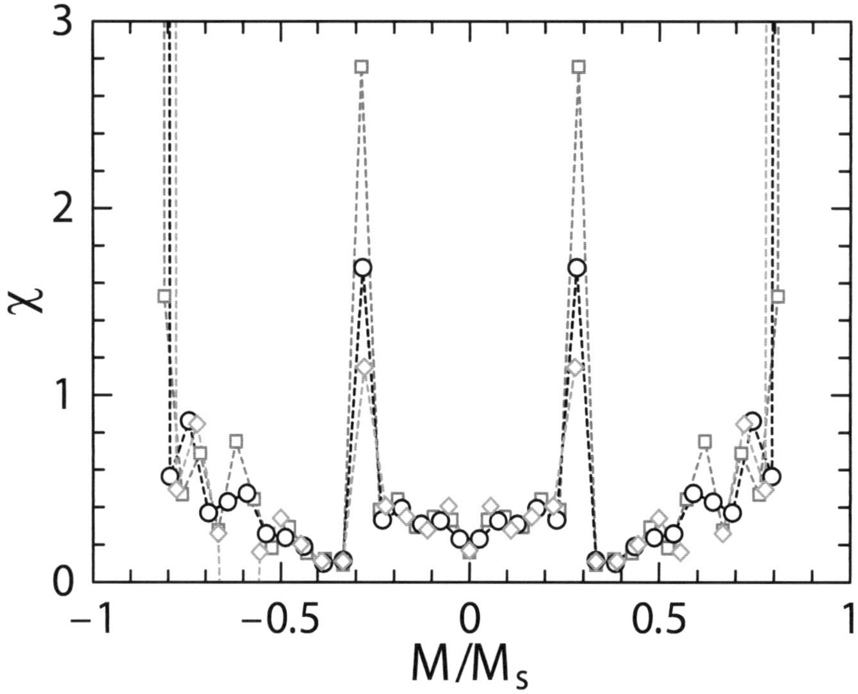

Figure 9. Magnetization dependence of χ of the kagome-lattice antiferromagnet for N =36, 39 and 42. The results for N =36, 39, and 42 are denoted by green diamonds, black circles, and red squares, respectively.

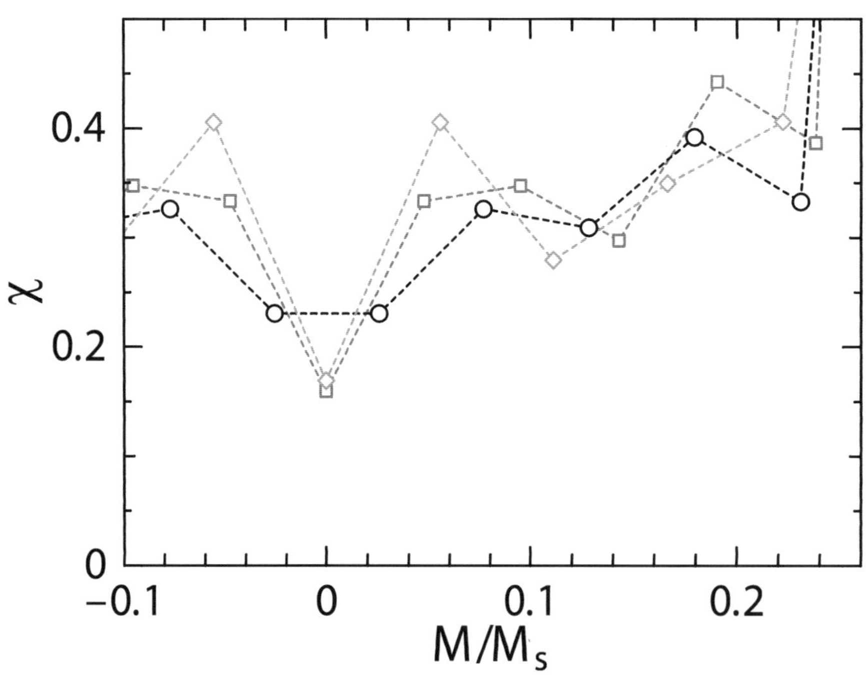

Figure 10. Magnetization dependence of χ of the kagome-lattice antiferromagnet for N =36, 39 and 42 around $m = 0$. Symbols are the same as in Fig. 9.

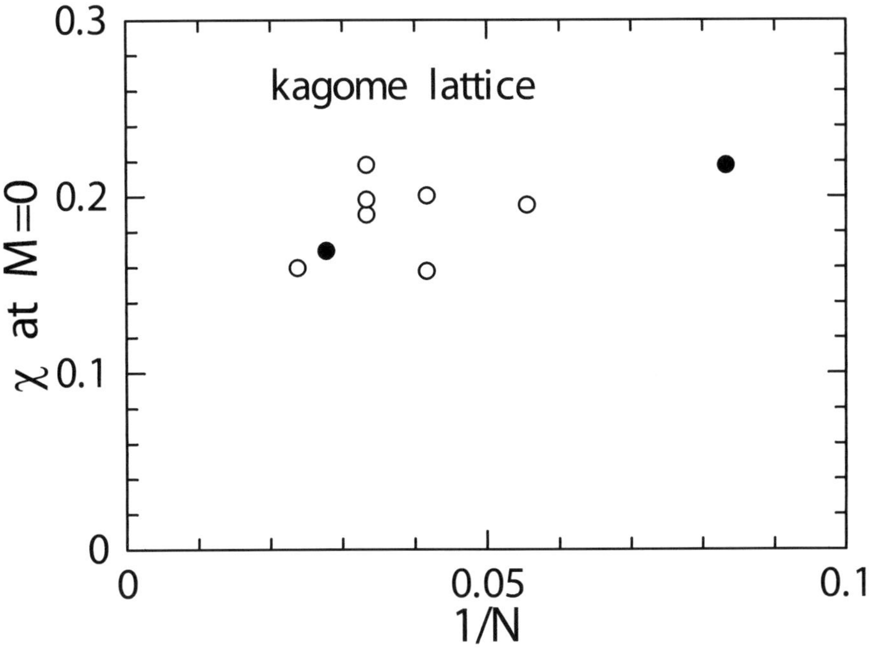

Figure 11. χ plotted versus $1/N$ for the kagome-lattice antiferromagnet. Closed symbols denote the results for the cases when the shape of the finite-size clusters are the rhombus with an interior angle $\pi/3$. When a finite-size cluster for a given N cannot form the same rhombus even if the rhombus is tilted, on the other hand, the cluster forms a parallelogram; the results are given by open symbols. The extrapolated value in the thermodynamic limit seems to be finite ($\chi \neq 0$).

Tohoku University; Supercomputing Division, Information Technology Center, The University of Tokyo for computational facilities. This work was partly supported by the Strategic Programs for Innovative Research, MEXT, and the Computational Materials Science Initiative, Japan. The authors would like to express their sincere thanks to the staff members of the Center for Computational Materials Science of the Institute for Materials Research, Tohoku University for their continuous support of the SR16000 supercomputing facilities.

References

[1] S. Sachdev, Phys. Rev. B **45**, 12377 (1992).
[2] T. Nakamura and S. Miyashita, Phys. Rev. B **52**, 9174 (1995).
[3] P. Lecheminant, B. Bernu, C. Lhuillier, L. Pierre, and P. Sindzingre, Phys. Rev. B **56**, 2521 (1997).
[4] Ch. Waldtmann, H.-U. Everts, B. Bernu, C. Lhuillier, P. Sindzingre, P. Lecheminant and L. Pierre, Eur. Phys. J. B **2**, 501 (1998).
[5] F. Mila, Phys. Rev. Lett. **81**, 2356 (1998).
[6] M. Hermele, T. Senthil and M. P. A. Fisher, Phys. Rev. B **72**, 104404 (2005).
[7] R. R. P. Singh and D. A. Huse, Phys. Rev. B **76**, 180407(R) (2007).
[8] Y. Ran, M. Hermele, P. A. Lee and X. -G. Wen, Phys. Rev. Lett. **98**, 117205 (2007).
[9] H. C. Jiang, Z. Y. Weng and D. N. Sheng, Phys. Rev. Lett. **98**, 117203 (2008).
[10] O. Cepas, C. M. Fong, P. W. Leung, and C. Lhuillier, Phys. Rev. B **78**, 140405(R) (2008).
[11] P. Sindzingre and C. Lhuillier, Europhys. Lett. **88**, 27009 (2009).
[12] G. Evenbly and G. Vidal, Phys. Rev. Lett. **104**, 187203 (2010).
[13] P. W. Anderson, Science **235**, 1196 (1987).
[14] M. P. Shores, E. A. Nytko, B. M. Barlett, and D. G. Nocera, J. Am. Chem. Soc. **127**, 13462 (2005).

[15] P. Mendels and F. Bert, J. Phys. Soc. Jpn. **79**, 011001 (2010).
[16] H. Yoshida, Y. Okamoto, T. Tayama, T.Sakakibara, M.Tokunaga, A. Matsuo, Y. Narumi, K. Kindo, M. Yoshida, M. Takigawa, and Z. Hiroi, J. Phys. Soc. Jpn. **78**, 043704 (2009).
[17] M. Yoshida, M. Takigawa, H. Yoshida, Y. Okamoto, and Z. Hiroi, Phys. Phys. Lett. **103**, 077207 (2009).
[18] Y. Okamoto, H. Yoshida, and Z. Hiroi, J. Phys. Soc. Jpn. **78**, 033701 (2009).
[19] H. Nakano and T. Sakai, J. Phys. Soc. Jpn. **80**, 053704 (2011).
[20] Y. Iqbal, D. Poilblanc and F. Becca, Phys. Rev. B **89**, 020407(R) (2014).
[21] Y. Iqbal, D. Poilblanc and F. Becca, Phys. Rev. B **91**, 020402(R) (2015).
[22] S. Yan, D. A. Huse and S. R. White, Science **332**, 1173 (2011).
[23] S. Depenbrock, I. P. McCulloch and U. Schollwöck, Phys. Rev. Lett. **109**, 067201 (2012).
[24] S. Nishimoto, N. Shibata and C. Hotta, Nat. Commun. **4**, 2287 (2013).
[25] S. Capponi, O. Derzhko, A. Honecker, A. M. Läuchli and J. Richter, Phys. Rev. B **88**, 144416 (2013).
[26] T. Han et al., Nature Commun. **492**, 406 (2012).
[27] T. Han et al., Phys. Rev. Lett. **108**, 157202 (2012).
[28] H. Nakano and A. Terai, J. Phys. Soc. Jpn. **78**, 014003 (2009).
[29] M. Matsumoto et al., Phys. Rev. B **65**, 014407 (2011).
[30] T. Sakai and H. Nakano, Phys. Rev. B Phys. Rev. B **83**, 100405(R) (2011).
[31] H. Nakano and T. Sakai, J. Phys. Soc. Jpn. **83**, 104710 (2014).

IOP Conf. Series: Journal of Physics: Conf. Series **868** (2017) 012007 doi:10.1088/1742-6596/868/1/012007

Quasi-two-dimensional magnetism in Cr-based MAX phases

Takeshi Waki, Zhongsheng Liu, Yoshikazu Tabata and Hiroyuki Nakamura

Department of Materials Science and Engineering, Kyoto University, Kyoto 606-8501, JAPAN

E-mail: `waki.takeshi.5c@kyoto-u.ac.jp`

Abstract. The MAX phase is one possible candidate for quasi-two-dimensional (Q2D) magnets owing to its layered structure. We report itinerant electron ferromagnetic and antiferromagnetic long-range orders in $(Cr_{1-x}Mn_x)_2GeC$ and Cr_2GaN, respectively. Because Cr_2GaN orders in a spin-density-wave state, the Q2D character is highly plausible. However, the Q2D nature is not straightforward in $(Cr_{1-x}Mn_x)_2GeC$ because of the absence of useful criteria to judge the dimensionality in the itinerant electron ferromagnet. In this article, we attempt to reveal the Q2D character of ferromagnetic $(Cr_{1-x}Mn_x)_2GeC$ by analyzing magnetic data in terms of the spin fluctuation theory for the Q2D system.

1. Introduction

Quasi-two-dimensional (Q2D) itinerant electron magnets have been attracting a great deal of attention in condensed matter physics research. Particularly, unconventional superconductivity in Q2D materials is one of the most familiar examples. A superconducting phase is often found in the vicinity of a magnetically ordered phase of layered materials such as cuprates and iron pnictides.

We have focused on the MAX phase as a new candidate for a Q2D itinerant electron magnet. The chemical formula of the MAX phase is $M_{n+1}AX_n$, where M is an early transition metal, A is a main group metal, X is carbon or nitrogen, and n = 1, 2, 3. The $M_{n+1}X_n$ layer and the A triangular lattice layer are piled up alternatively to form the layered structure [1]. Although the known stable phases were found in the 1960s–70s [2], the magnetism of the MAX phase has remained almost unexplored. Among stable MAX phases, only Cr-based materials are likely to be magnetic. We rst surveyed the magnetism of nondoped Cr-base materials Cr_2AlC, Cr_2GaC, Cr_2GeC, and Cr_2GaN and found that the carbides are all Pauli paramagnetic, whereas Cr_2GaN shows a spin-density-wave (SDW) transition at 170 K [3]. As far as we know this is the only example of the bulk nondoped magnetically ordered MAX phase. However, room temperature ferromagnetism in a $(Cr_{0.75}Mn_{0.25})_2GeC$ thin lm has been reported [4]. Independently, we con rmed bulk ferromagnetism and its systematic enhancement in the $(Cr_{1-x}Mn_x)_2GeC$ system [5].

The Q2D magnetism in localized spin systems is often discussed based on the magnetic susceptibility . Deviation from the Curie-Weiss law and a broad maximum in at a low temperature associated with magnetic short-range order are signatures of Q2D magnetism. In the itinerant electron case, the appearance of the SDW state suggests strongly Q2D

IOP Conf. Series: Journal of Physics: Conf. Series **868** (2017) 012007 doi:10.1088/1742-6596/868/1/012007

antiferromagnetism because it often originates from the low-dimensional nature of the Fermi surfaces and their nesting. This is the case of Cr_2GaN. However, it is di cult to discuss the Q2D character in $(Cr_{1-x}Mn_x)_2GeC$ because of the absence of established criteria for the itinerant electron ferromagnet.

Spin fluctuation parameters, which can be estimated from macroscopic magnetic data using Takahashi's spin fluctuation theory [6], would be helpful to discuss Q2D itinerant electron ferromagnetism. Recently, spin fluctuation parameters were estimated for itinerant electron ferromagnets with layered structures [7, 8, 9, 10, 11, 12]. In the generalized Rhodes-Wohlfarth plot, some systems exhibit signi cant deviation from the three-dimensional (3D) theoretical relation. However, it is di cult to judge whether the deviation is due to low dimensionality.

In this article, as a more quantitative discussion for the systematic deviation from the 3D line for $(Cr_{1-x}Mn_x)_2GeC$, we analyzed macroscopic magnetic data using Takahashi's spin fluctuation theory [6] and its extension to the Q2D system [13].

2. Experiments

Polycrystalline samples of $(Cr_{1-x}Mn_x)_2GeC$ were synthesized by solid state reaction of powder elements in evacuated quartz silica tubes. The heating temperature was 950–1000°C. Macroscopic magnetic measurements were performed using a SQUID magnetometer (MPMS, Quantum Design) at T = 2–300 K and H = 0–7 T.

3. Analysis

Here we describe the analysis used to estimate spin fluctuation parameters and the index of dimensionality using magnetization data.

Wave number and frequency dependencies of the imaginary part of the dynamical susceptibility for weak itinerant electron ferromagnets can be described by a double Lorentzian spectrum. As measures of spectral widths in wave-vector and energy spaces, T_A and T_0 are introduced, respectively [14].

On the other hand, Landau's expansion of the magnetic free energy at the ground state is described as

$$F_{\mathrm{m}}(M) = F_{\mathrm{m}}(0) + \frac{1}{2(g\mu_{\mathrm{B}})^2}M^2 + \frac{\bar{F}_1}{4(g\mu_{\mathrm{B}})^4 N_0^3}M^4, \tag{1}$$

where g is the Lande's g-factor, is the magnetic susceptibility, μ_{B} is the Bohr magneton and N_0 is the number of magnetic ions. $\bar{F}_1$ corresponds to the slope of the Arrott plot. Experimentally we can estimate $\bar{F}_1$ with the relation as

$$\bar{F}_1 = \frac{N_0^3(2\mu_{\mathrm{B}})^4}{\xi k_{\mathrm{B}}}, \tag{2}$$

where ξ is the slope of the Arrott plot at zero temperature and k_{B} is the Boltzmann constant.

In Takahashi's spin fluctuation theory, $\bar{F}_1$ is related to T_A and T_0 as

$$\bar{F}_1 = \frac{4T_A^2}{15T_0}. \tag{3}$$

We also obtain the relation among the spontaneous moment at the ground state, p_{s}, T_A and T_0 as

$$p_{\mathrm{s}}^2 = \frac{20T_0}{T_A}C_{4/3}\ \frac{T_{\mathrm{C}}}{T_0}^{4/3}, \tag{4}$$

where $C_{4/3}$=1.00608 . Using Eqs.(3) and (4), we can estimate T_A and T_0 from macroscopic magnetic data.

Takahashi's spin fluctuation theory is extended to the Q2D system by introducing the anisotropy to the electron e ective mass. As a measure of the dimensionality, $^2 = m/m'$ is used, where m and m' are the e ective in-plane and out-of-plane components of the mass, respectively. In this treatment, the relation among $\bar{F}_1$, T_A and T_0 can be modi ed from the 3D case as

$$\bar{F}_1 = \frac{2T_A^2}{5d_z T_0}, \qquad d_z = \frac{6 \quad 3}{3 \quad 2}. \tag{5}$$

As a result, we obtain, for the Q2D case, the relations

$$\frac{T_C}{T_A}^{5/3} = \frac{p_s^2}{5g^2 C_{4/3}} \left(\frac{5d_z \bar{F}_1}{2T_C}\right)^{-1/3} \tag{6}$$

and

$$\frac{T_C}{T_0}^{5/6} = \frac{p_s^2}{5g^2 C_{4/3}} \left(\frac{5d_z \bar{F}_1}{2T_C}\right)^{1/2} \tag{7}$$

which enable us to calculate the spin fluctuation parameters using macroscopic data. When we calculate T_A and T_0 using these equations, should be set in addition to macroscopic magnetic data. To estimate , we use the generalized Rhodes-Wohlfarth plot.

Although the original Rhodes-Wohlfarth plot was proposed empirically, Takahashi showed that p_{eff}/p_s plotted against T_C/T_0 satis es

$$\frac{p_{\text{eff}}}{p_s} = 1.4 \quad \frac{T_C}{T_0}^{-2/3} \tag{8}$$

for a 3D itinerant electron weak ferromagnet. For the Q2D case, the relation between p_{eff}/p_s and $t_c = T_C/T_0$ is given as a function of as

$$\frac{p_{\text{eff}}}{p_s}^{2} = \frac{3}{10 d_T t_c T(t_c, \ )} \quad \frac{dy}{dt}^{-1}, \tag{9}$$

with

$$d_T = \frac{6}{3 \quad 2}, \quad t_c T(t_c, \ ) = {}^3\big(H_{3D}(0, t_c, \ ) + H_{2D}(0, t_c, \ )\big), \tag{10}$$

$$y = \frac{N_0(g\mu_B)^2}{2k_B T_A}, \tag{11}$$

$$y = \frac{3}{1 \quad /2} \quad \big(H_{3D}(y, t, \ ) + H_{2D}(y, t, \ )\big) \quad \big(H_{3D}(0, t_c, \ ) + H_{2D}(0, t_c, \ )\big) \ , \tag{12}$$

$$H_{3D}(y, t, \ ) = \int_0^1 z^3 dz \quad \ln \nu \quad \frac{1}{2\nu} \quad \psi(\nu) \ , \tag{13}$$

$$H_{2D}(y, t, \ ) = \int_1^{1/\epsilon} z^2 dz \quad \ln \nu \quad \frac{1}{2\nu} \quad \psi(\nu) \ , \tag{14}$$

$$t = \frac{T}{T_0}, \quad \eta = y/ \ ^2, \quad \tau = t/ \ ^3, \quad \nu = z(\eta + z^2)/\tau, \tag{15}$$

where $\psi(\nu)$ is the digamma function. In contrast ot the 3D case, the theoretical relation for the Q2D case needs numerical calculations, because Eq.(9) is not analytic. Calculated curves deviate downward from the 3D line (Eq.(8)) with decreasing . By using these curves we can estimate for the data points. Since the location of the data points in the plot depends on the presumed value to calculate T_0, should be determined self-consistently.

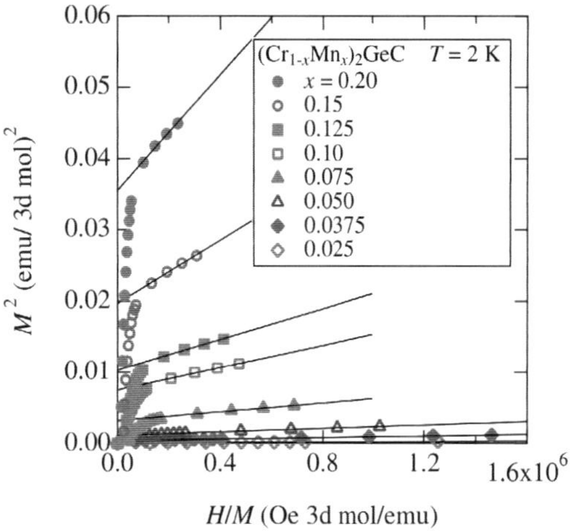

Figure 1. Arrott plot for $(\mathrm{Cr}_{1-x}\mathrm{Mn}_x)_2\mathrm{GeC}$ at $T = 2$ K. Solid lines indicate the best ts for di erent concentration data.

4. Results and Analysis

Figure 1 shows the Arrott plot for $(\mathrm{Cr}_{1-x}\mathrm{Mn}_x)_2\mathrm{GeC}$ at $T = 2$ K. Since the T_C values of $(\mathrm{Cr}_{1-x}\mathrm{Mn}_x)_2\mathrm{GeC}$ are su ciently high compared to 2 K, we estimate p_s and F_1 from these data by regarding the magnetization as that of the ground state. Solid lines are the best ts of the data to

$$M^2 = p_\mathrm{s}^2 + \xi \frac{H}{M}. \tag{16}$$

With increasing x, the M^2-intercept or p_s increases monotonically. The slope also increases, suggesting that changes in the spin fluctuation are associated with the Mn substitution.

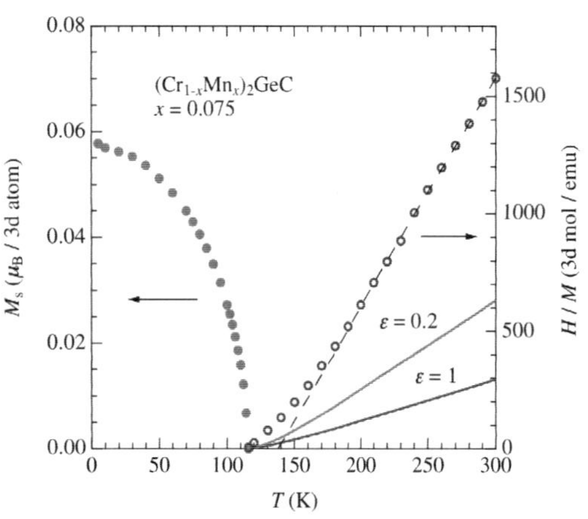

Figure 2. Temperature dependence of magnetization and reciprocal susceptibility for $x = 0.075$ estimated from the Arrott plot. The solid circles represent the temperature dependence of $M_\mathrm{s}(T)$ (left axis). Open circles represent reciprocal susceptibility (right axis). The dashed line is the best t to the Curie-Weiss law. Solid curves represent calculated reciprocal susceptibility for $= 0.2$ and 1.

In Fig. 2, the temperature dependence of the spontaneous moment $M_\mathrm{s}(T)$ estimated from the Arrott plot is shown for $x = 0.075$. With increasing temperature, $M_\mathrm{s}(T)$ decreases monotonically to be zero at $T_\mathrm{C} = 115$ K. The magnetic susceptibility estimated from the Arrott plot was tted with the modi ed Curie-Weiss law

$$(T) = \frac{N_0 p_\mathrm{eff}^2 \mu_\mathrm{B}^2}{3k_\mathrm{B}(T \quad \theta)} + \ _0, \tag{17}$$

where θ and $_0$ are the Weiss temperature and temperature independent susceptibility, respectively. The obtained parameters are $p_\mathrm{eff} = 0.888\ \mu_\mathrm{B}$, $\theta = 138$ K, and $_0 = 0.00019$

emu/mol. We show the temperature dependent term of the reciprocal susceptibility to compare it to the theoretical curves.

Table 1. Magnetic data and spin fluctuation parameters for $(Cr_{1-x}Mn_x)_2GeC$. To calculate T_A-Q2D and T_0-Q2D, we used the Q2D index $= 0.2$. T_0-Q2D was used to calculate T_C/T_0.

x	T_C (K)	p_{eff} (μ_B)	p_s (μ_B)	$\bar{F}_1$ (10^4 K)	T_A-3D (10^4 K)	T_A-Q2D (10^4 K)	T_0-3D (10^4 K)	T_0-Q2D (10^4 K)	p_{eff}/p_s	T_C/T_0 10^{-3}
0.025	43	0.456	0.00961	831	102	106	3.31	2.94	47.5	1.46
0.0375	48	0.596	0.0197	201	35.2	36.6	1.62	1.44	30.2	3.30
0.050	75	0.712	0.0335	88.2	22.6	23.5	1.54	1.37	21.2	5.47
0.075	115	0.888	0.0566	35.2	14.1	14.7	1.51	1.34	15.7	8.56
0.10	150	1.15	0.0867	13.8	8.68	9.02	1.46	1.30	13.3	11.6
0.125	193	1.18	0.101	9.99	8.25	8.58	1.82	1.62	11.6	11.9
0.15	205	1.33	0.140	4.84	5.08	5.28	1.42	1.26	9.47	16.2
0.20	270	1.47	0.189	2.65	3.93	4.09	1.55	1.38	7.82	19.5

The magnetic data obtained are listed in Table 1. When we assume a 3D nature, T_A and T_0 can be readily calculated from these values. Even if we assume a 3D nature, the data for $(Cr_{1-x}Mn_x)_2GeC$ plotted in the form of $\log(p_{eff}/p_s)$ vs. $\log(T_C/T_0)$ are systematically shifted downward from the 3D line [5].

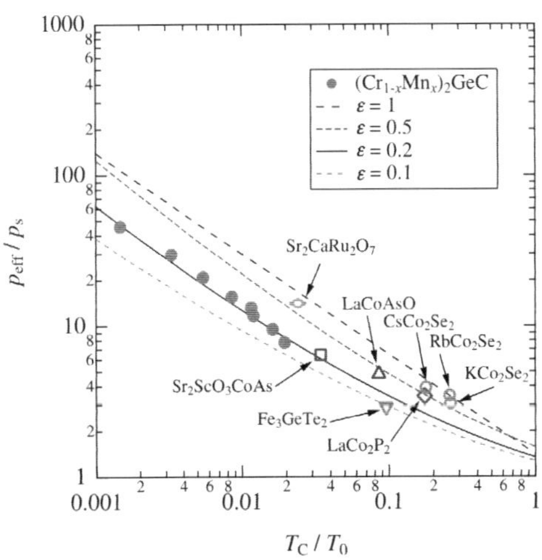

Figure 3. Generalized Rhodes-Wohlfarth plot for $(Cr_{1-x}Mn_x)_2GeC$. Theoretical curves for 3D and Q2D cases are drawn. Data points for $Sr_2CaRu_2O_7$ [7], LaCoAsO [8], Fe_3GeTe_2 [9], Sr_2ScO_3CoAs [10] ACo_2Se_2 (A = K, Rb, Cs) [11] and $LaCo_2P_2$ [12] are also included.

If we take into account , the data points slightly move to the right from the 3D case. As shown in Fig. 3, all the experimental data agree well with the case for $= 0.2$, suggesting that $(Cr_{1-x}Mn_x)_2GeC$ are Q2D itinerant electron weak ferromagnets and that the dimensionality of this series is insensitive to the Mn content. The latter is consistent with the fact that lattice constants are almost independent of x [5]. The obtained value implies that the out-of-plane component of the e ective mass is about 25 times larger than that of the in-plane one. To check the validity of the analysis, experimental estimation of the e ective mass would be helpful.

In the gure, we add the data taken from the literature for other itinerant electron weak ferromagnets with layered structures. Although is not included when T_0 is estimated for these materials, some of the data points show an apparent downward deviation from the 3D line, suggesting a Q2D character. This plot would be useful to estimate the dimensionality for itinerant electron ferromagnets.

Using Eqs.(11) and (12), we calculated the reciprocal susceptibility for $x = 0.075$ for 3D ($T_A = 14.1 \times 10^4$ K, $T_0 = 1.51 \times 10^4$ K and $= 1$) and Q2D ($T_A = 14.7 \times 10^4$ K, $T_0 = 1.34 \times 10^4$ K and $= 0.2$) cases, which are shown in Fig. 2. The reciprocal susceptibility calculated for the 3D case is about 1/4 that of the experimental one. By taking into account the Q2D nature, the agreement between theory and experiments is improved but not satisfactorily. For further improvement of the theory, Q2D analyses for other layered systems are desired. We do not dare to estimate the spin fluctuation parameters and from the reciprocal susceptibility because they cannot be determined independently.

5. Summary

We analyzed macroscopic magnetic data of $(Cr_{1-x}Mn_x)_2GeC$ using the spin fluctuation theory for Q2D itinerant electron ferromagnets. We observed a systematic deviation from the theoretical 3D line in the generalized Rhodes-Wohlfarth plot, whose appearance indicates a Q2D character. The index of dimensionality for the $(Cr_{1-x}Mn_x)_2GeC$ series was estimated to be $= 0.2$ from a comparison of the data points and the theoretical curve in the plot.

References

[1] Barsoum M W 2000 *Prog. Solid State Chem.* **28** 201–281
[2] Nowotny V H 1971 *Progr. Solid State Chem.* **5** 27–70
[3] Liu Z, Waki T, Tabata Y, Yuge K, Nakamura H and Watanabe I 2013 *Phys. Rev.* B **88** 134401
[4] Ingason A S, Mockuté A, Dahlqvist M, Magnus F, Olafsson S, Arnalds U B, Alling B, Abrikosov I A, Hjorvarsson B, Persson P Å and Rosen J 2013 *Phys. Rev. Lett.* **110** 195502
[5] Liu Z, Waki T, Tabata Y and Nakamura H 2014 *Phys. Rev.* B **89** 054435
[6] Takahashi Y 1986 *J. Phys. Soc. Jpn.* **55** 3553–3573
[7] Ikeda S, Maeno Y and Fujita T 1998 *Phys. Rev.* B **57** 978
[8] Ohta H and Yoshimura K 2009 *Phys. Rev.* B **79** 184407
[9] Chen B, Yang J, Wang H, Imai M, Ohta H, Michioka C, Yoshimura K and Fang M 2013 *J. Phys. Soc. Jpn.* **82** 124711
[10] Ohta H, Noguchi D, Nabetani K and Katori H A 2013 *Phys. Rev.* B **88** 094441
[11] Yang J, Chen B, Wang H, Mao Q, Imai M, Yoshimura K and Fang M 2013 *Phys. Rev.* B **88** 064406
[12] Imai M, Michioka C, Ueda H and Yoshimura K 2015 *Phys. Rev.* B **91** 184414
[13] Takahashi Y 1997 *J. Phys.: Condens. Matter* **9** 10359
[14] Takahashi Y and Moriya T 1985 *J. Phys. Soc. Jpn.* **54** 1592–1598

IOP Conf. Series: Journal of Physics: Conf. Series **868** (2017) 012008 doi:10.1088/1742-6596/868/1/012008

Electronic states in the pressure-induced magnetically ordered phase in SmB_6

Takeshi Mito[1], Naoya Emi[1], Naomi Kawamura[2], Masaichiro Mizumaki[2], Takehide Koyama[1], Koichi Ueda[1], Naoki Ishimatsu[3], and Fumitoshi Iga[4]

[1] Graduate School of Material Science, University of Hyogo, Kamigori, Hyogo 678-1297, Japan
[2] Japan Synchrotron Radiation Research Institute, SPring-8, Sayo, Hyogo 679-5198, Japan
[3] Graduate School of Science, Hiroshima University, Hiroshima 739-8526, Japan
[4] Faculty of Science, Ibaraki University, Mito, Ibaraki 310-8512, Japan

E-mail: mito@sci.u-hyogo.ac.jp

Abstract. We have carried out the high-pressure measurement of X-ray absorption spectroscopy on the intermediate valence compound SmB_6 which shows magnetic ordering as well as an insulator-metal transition at critical pressure $P_c \sim 10$ GPa. The valence of Sm atom at room temperature increases with increasing pressure, however it is far below a trivalent state at P_c. In contrast to cases of pressure-induced nonmagnetic-magnetic transition in Yb compounds, which mostly occurs in the scheme of well localized $4f$ electrons, the present observation suggests that electronic system in SmB_6 still possesses strong delocalized characters at P_c.

1. Introduction

Rare-earth based compounds show variety of ground states. The ground states are often varied to another by changing pressure or magnetic field, resulting in for example a transition between nonmagnetic and magnetic states, between metallic and semiconducting states, and so on. In Ce, Sm, Eu, and Yb compounds, the valence of rare-earth ions plays a key role to control their features. According to the ground state of Hund's rule, trivalent (divalent for Eu) states show magnetic features, while divalent (trivalent for Eu and tetravalent for Ce) states are nonmagnetic. Therefore a change in the valence intimately relates to a transition between nonmagnetic and magnetic ground states in these compounds.

Pressure-induced nonmagnetic-magnetic transitions in Yb compounds have been investigated so far. In the Yb systems, pressure tends to induce the localization of $4f$ electrons from the so-called intermediate valence state. The X-ray absorption spectroscopy (XAS) is a powerful technique to measure the mean valence of the rare-earth ions even under high pressure. According to the results of the high-pressure XAS measurements performed on various Yb compounds (for example, see Ref. [1, 2, 3]), most of the pressure-induced magnetic orders appear with mean Yb valence close to 3+, namely in the scheme of well localized $4f$ electrons.

It is also known that some Sm compounds undergo a nonmagnetic-magnetic transition under high pressure. Therefore, it is useful to compare the effects of pressure on the two different systems in order to study the nonmagnetic-magnetic transition in the rare-earth compounds systematically. SmB_6 is a well-known intermediate valence compound having a

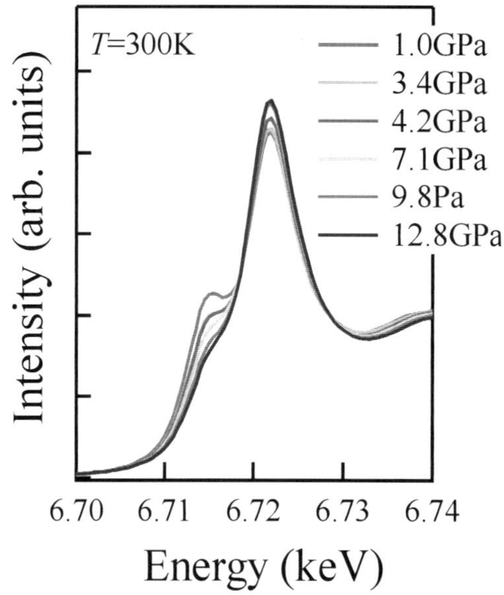

Figure 1. Sm L_3-edge absorption spectra of SmB_6 measured at 300 K and at 1.0, 3.4, 4.2, 7.1, 9.8, and 12.8 GPa.

small semiconducting gap of 50 – 100 K [4, 5]. The formation of the gap may be associated with hybridization between conduction and $4f$ electrons, however the details are still controversial. Recent nuclear forward scattering, specific heat, and resistivity measurements under pressure in SmB_6 give evidences for the appearance of magnetic ordering simultaneously with an insulator-metal transition at critical pressure $P_c = 7 - 10$ GPa [6, 7, 8]. No structural anomaly has been found so far under pressure [13]. However, the detailed electronic states at the transition, especially whether $4f$ electrons are in a well localized state just like in many Yb compounds, have not been clarified yet. We have therefore carried out XAS measurements in order to investigate the pressure dependence of Sm valence in SmB_6.

2. Experimental details

Single crystalline samples of SmB_6 were grown by a floating-zone method using an image furnace with four xenon lamps [9]. The XAS measurements near the Sm L_3-edge (6.72 keV) were performed in the pressure range of 1 to 13 GPa at room temperature, at the beamline BL39XU of SPring-8, Japan [10]. For the high pressure measurements, the sample was loaded in a diamond anvil cell (DAC) filled with a mixture of 4 : 1 methanol : ethanol as a pressure-transmitting medium. Nanopolycrystalline diamond anvils were used to avoid glitches in XAS spectra [11]. The pressure was calibrated using fluorescence from ruby chips mounted with the sample inside the DAC.

3. Results and Discussion

Figure 1 shows the Sm L_3-edge absorption spectra of SmB_6. The shoulder like structure at 6.715 keV, corresponding to a divalent component, is gradually suppressed with increasing pressure. However, a finite spectral weight persists even at the maximum pressure of the present experiment. Simultaneously, the intensity of the main peak at 6.72 keV, corresponding to a trivalent component, increases with pressure. These results clearly indicate that pressure brings

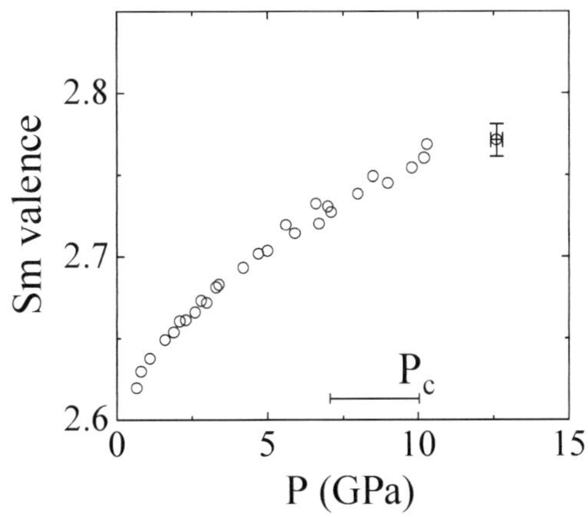

Figure 2. Estimated mean Sm valence at 300 K as a function of pressure. P_c somewhat distributes depending on the experiments [6, 7, 8].

about a shift of the Sm valence toward the trivalent state, consistent with the results of previous Raman scattering and nuclear magnetic resonance measurements [12, 13].

The mean valence of the Sm ions is estimated from the relative intensities of the Sm^{2+} and Sm^{3+} components in the XAS spectra shown in Fig. 1. Each component was modeled by the sum of a Lorentz function and an arctangent function representing the continuum excitations. The evaluated Sm valence v_{Sm} is illustrated in Fig. 2 as a function of pressure. As pressure increases, v_{Sm} increases and the slope dv_{Sm}/dP gradually decreases. We here compare the present observations with various cases of pressure-induced nonmagnetic-magnetic transition in the Yb compounds (for example, see Ref. [1, 2, 3]). The weak saturation of v_{Sm} at high pressures in SmB_6 is analogous to the pressure dependence of Yb valence in the Yb compounds. However, a striking contrast between the two systems is observed in the valence at the critical pressure: at the nonmagnetic-magnetic transition in the Yb compounds, the mean Yb valence is very close to 3+, while v_{Sm} in SmB_6 is far below 3+, *i.e.* $v_{\mathrm{Sm}} < 2.8$ at P_c. The result indicates that electronic system in SmB_6 still possesses strong delocalized characters at P_c. The remarkable difference in the valence of the rare-earth atoms should arise from underlying electronic structures and correlations unique to each rare-earth compound at the critical pressure.

4. Summary

We have carried out XAS measurements on the intermediate compound SmB_6 in the pressure range up to 13 GPa at room temperature. The Sm valence increases with increasing pressure, however the valence is far below the trivalent state at the critical pressure where magnetic ordering occurs. With respect to the mean valence of rare-earth atoms, the present observation in SmB_6 strikingly contrasts with what is mostly observed in the Yb compounds: namely the Yb valence is very close to 3+ at critical pressure. The result suggests that electronic system in SmB_6 still possesses strong delocalized characters at P_c.

5. Acknowledgment

The authors are grateful to H. Sumiya and T. Irifune for providing the nano-polycrystalline diamond anvils. This work was supported by JSPS KAKENHI (Grant No.24540349). The synchrotron radiation experiments were performed at the BL39XU of SPring-8 with the approval of the Japan Synchrotron Radiation Research Institute (JASRI) (Proposal Nos. 2014A1233,

2014B1564, and 2014B2041).

References

[1] Yamaoka H, Jarrige I, Tsujii N, Lin J-F, Hiraoka N, Ishii H and Tsuei K-D 2010 *Phys. Rev. B* **82** 035111
[2] Fernandez-Pañella A, Balédent V, Braithwaite D, Paolasini L, Verbeni R, Lapertot G and Rueff J.-P 2012 *Phys. Rev. B* **86** 125104
[3] Yamaoka H, Tsujii N, Utsumi Y, Sato H, Jarrige I, Yamamoto Y, Lin J-F, Hiraoka N, Ishii H, Tsuei K-D and Mizuki J 2013 *Phys. Rev. B* **87** 205120
[4] Kasuya T, Takegahara K, Fujita T, Tanaka T and Bannai E 1979 *J. Phys. Colloq.z* **40** C5-308
[5] Takigawa M, Yasuoka H, Kitaoka Y, Tanaka T, Nozaki H and Ishizawa Y 1981 *J. Phys. Soc. Jpn.* **50** 2525
[6] Barla A, Derr J, Sanchez J P, Salce B, Lapertot G, Doyle B P, Rüffer R, Lengsdorf R, Abd-Elmeguid M M and Flouquet J 2005 *Phys. Rev. Lett.* **94** 166401
[7] Derr J, Knebel G, Braithwaite D, Salce B, Flouquet J, Flachbart K, Gabáni S and Shitsevalova N 2008 *Phys. Rev. B* **77** 193107
[8] Derr J, Knebel G, Lapertot G, Salce B, Méasson M-A and Flouquet J 2006 *J. Phys.: Condens. Matter* **18** 2089
[9] Iga F, Shimizu N and Takabatake T 1998 *J. Magn. Magn. Mater.* **177-181** 337
[10] Kawamura N, Ishimatsu N and Maruyama H 2009 *J. Synchrotron Radiat.* **16** 730.
[11] Ishimatsu N, Matsumoto K, Maruyama H, Kawamura N, Mizumaki M, Sumiya H and Irifune T, 2012 *J. Synchrotron Radiat.* **19** 768.
[12] Ogita N, Nagai S, Udagawa M, Iga F, Sera M, Oguchi T, Akimitsu J and Kunii S 2005 *Physica B* **359-361** 941
[13] Nishiyama K, Mito T, Pristáš G, Hara Y, Koyama T, Ueda K, Kohara T, Akahama Y, Gabáni S, Reiffers M, Flachbart K, Fukazawa H, Kohori Y, Takeshita N and Shitsevalova N 2013 *J. Phys. Soc. Jpn.* **82** 123707

IOP Conf. Series: Journal of Physics: Conf. Series **868** (2017) 012009 doi:10.1088/1742-6596/868/1/012009

Magnetic study of $FeSi_{1-x}Ge_x$ across the nonmagnetic-ferromagnetic transition

Naohito Tsujii

National Institute for Materials Science, 1-2-1 Sengen, Tsukuba 305-0047, Japan

E-mail: tsujii.naohito@nims.go.jp

Abstract. Magnetic properties of $FeSi_{1-x}Ge_x$ have been investigated using polycrystalline samples. A transition from nonmagnetic state with an energy gap for $x = 0$ to a weak ferromagnetic state for $x = 0.5$ has been confirmed, which was previously reported by Yeo et al. using single crystalline samples. Pauli-paramagnetic contribution to the low temperature susceptibility, $\chi_{\rm spin}(0)$, has been estimated for the samples with $x \leq 0.3$ by fitting the observed data. It significantly increases with the increase of x, reaching the value $\chi_{\rm spin}(0) = 0.75\times10^{-3}$ emu/mol, which is comparable to those of the nearly ferromagnetic metals YCo_2 and Pd. Using the Arrot plot, the parameter F_1 of the spin-fluctuation theory has been estimated. F_1 shows a drastic decrease from the order of 10^6 K for $x = 0$ to 9.5×10^3 K for $x = 0.5$, which is very similar to the x dependence of F_1 in $Y(Co_{1-x}Al_x)_2$ across the nonmagnetic-ferromagnetic phase boundary. These results suggest that the evolution of the magnetism in $FeSi_{1-x}Ge_x$ can be well understood in terms of the transition from nearly ferromagnetic to weak itinerant ferromagnetic metal.

1. Introduction

FeSi crystallizes in the cubic B20-type structure. Physical properties of FeSi have extensively been studied over several decades because of its unusual magnetic behavior. The ground state is nonmagnetic semiconductor with an energy gap of about 60 meV, as directly observed by the photoelectron spectroscopy measurement [1]. Upon heating, the magnetic susceptibility increases rapidly and exhibits a maximum at 500 K [2]. Above the maximum, the magnetic susceptibility decreases with temperature, showing a Curie-Weiss like behavior. Jaccarino et al. first explained this unusual magnetic behavior in terms of an activation behavior, but they had to assume unrealistically sharp band structures [2]. Later, Moriya and Takahashi suggested based on their self-consistent renormalization (SCR) theory, that the anomalous magnetic behavior in FeSi is explained by temperature-induced magnetic moments, which arises through thermally excited spin-fluctuations [3, 4, 5]. The existence of temperature-induced moments was indeed confirmed by neutron scattering experiments [6, 7].

In the SCR theory, a negative mode-mode coupling constant was assumed to explain the magnetism of FeSi. This should lead a negative constant F_1, which corresponds to the fourth-order expanded coefficient of the free energy in magnetization. However, recent magnetization measurement using a single crystalline FeSi sample revealed that the coefficient F_1 is actually positive and show an exponential increase at low temperatures [8]. Very similar temperature dependence of F_1 has also been observed in the semiconductor $FeSb_2$ [9]. Takahashi has given a clue to solve this discrepancy by taking the effect of quantum spin-fluctuation into consideration [10, 11]. The theory of Takahashi explained both the temperature dependence of the magnetic susceptibility and the positively increasing F_1 of FeSi. Hence, the magnetism of FeSi has well been explained within the framework of itinerant electron magnetism with enhanced spin fluctuations.

On the other hand, another model based on the localized moment picture was proposed [12, 13]. They claimed that the temperature variation of the magnetic and electrical properties of FeSi can be explained by the Kondo insulator formation. Schlesinger et al. pointed through the optical conductivity measurement that the energy gap of FeSi disappears much faster at elevated temperatures than expected from its gap size, and suggested that this damping can be explained within the framework of a Kondo insulator picture [14]. However, resonant X-ray inelastic scattering experiments did not detect any signature of valence admixture in the Fe 3d state over different temperatures [15, 16]. This result is in favor of itinerant electron model.

Magnetic properties of the solid solution system $FeSi_{1-x}Ge_x$ is also intriguing. Bauer et al. showed that the energy gap of FeSi becomes narrower by substituting Ge for Si, and a metallic state evolves at $x = 0.5$ [17]. This is consistent with the increase of the gap size of FeSi under pressure [17, 18], since alloying with Ge acts as a negative pressure on FeSi. Later, a local density approximation calculation suggested that a first-order phase transition from a nonmagnetic insulator to a ferromagnetic metal occurs in $FeSi_{1-x}Ge_x$ at around $x =$ 0.4 [19]. This phase transition was indeed observed experimentally at $x_c \simeq 0.3$ by Yeo et al. [20] Furthermore, they revealed that the electronic specific heat coefficient γ shows a sharp peak at x_c, suggesting the evolution of strongly correlated electronic state around this chemical composition. This phase transition is quite unique since usual metal-insulator transitions occur from a nonmagnetic metal to a magnetically-ordered insulator. In addition, the compounds near x_c is likely to be a semiconductor with strong electronic correlation.

This motivated us to start the research of the electronic properties of $FeSi_{1-x}Ge_x$, since such correlated-electron state can enhance various electronic properties such as thermoelectric power and heat capacity. In this papaer, we first focus on the sample preparation, characterization, and the magnetic properties. We will show by our magnetization results that the phase transition in $FeSi_{1-x}Ge_x$ is well understood in terms of itinerant electron system. Thermal properties of the same samples will be presented as another paper.

2. Experimental

Polycrystalline samples of $FeSi_{1-x}Ge_x$ were synthesized by means of arc melting and subsequent annealing. Pure elements of Fe (4N), Si (11N), and Ge (6N) were weighted and placed on a water-cooled Cu hearth, and arc-melted under an argon atmosphere. Ge was found to be volatile during the melting process. Therefore, extra amounts of Ge were loaded to compensate the loss. The weight loss after arc-melting was less than 1 % of the total mass. Hereafter we refer these arc-melted samples to as-cast samples. The as-cast ingots were sealed in quartz tubes under vacuum and annealed for several days. Samples of $FeSi_{1-x}Ge_x$ with $x = 0$ to 0.22 were annealed at 1227 K. On the other hand, the sample with $x = 0.3$ was found to be partially melted when rapidly heated up to 1273 K. Therefore, the $x = 0.3$ sample was first annealed at 1173 K for one day, then heated up to 1273 K and annealed for 3 days. The sample for $x = 0.5$ was annealed at 1173 K, above which the sample was found to be partially melted.

Powder X-ray diffraction (XRD) was collected using the Cu Kα radiation. The XRD patterns were analyzed by the Rietveld method using the FullProf software [21]. Magnetic susceptibility and magnetization were measured using a superconducting quantum interference device (SQUID) magnetometer, MPMS-5S, provided from Quantum Design.

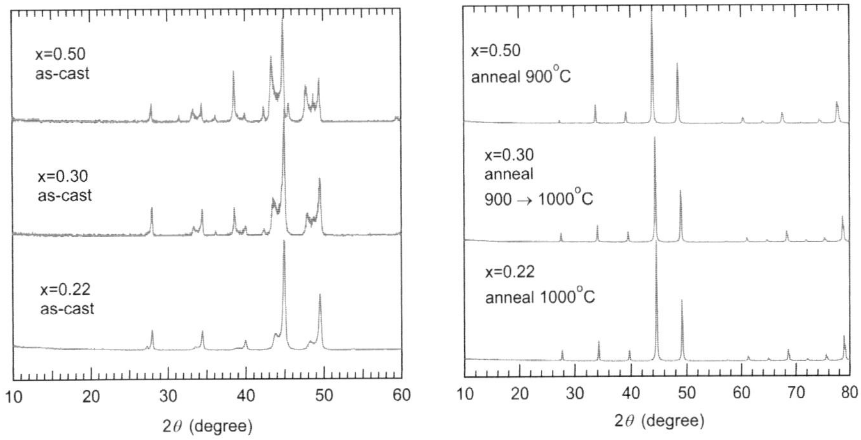

Figure 1. Powder X-ray diffraction patterns of $FeSi_{1-x}Ge_x$ samples. Left panel shows the patterns of the as-cast samples. Right panel shows those of the annealed samples.

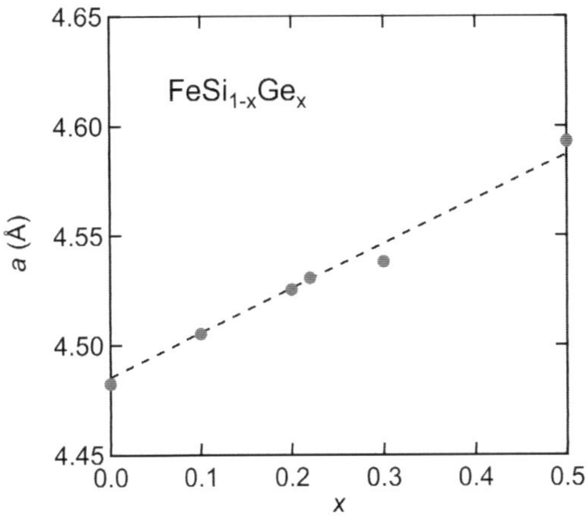

Figure 2. Lattice constants of the $FeSi_{1-x}Ge_x$ annealed samples as a function of x. Broken line shows the Vegard's relation estimated from $a = 4.4853$ Å of FeSi [8] and $a = 4.689$ Å of FeGe [22].

3. Results and discussion

In Figure 1, powder x-ray diffraction patterns of the as-cast and the annealed $FeSi_{1-x}Ge_x$ samples are shown. It is found that the as-cast samples show split peaks, indicating that the samples contain multiple phases with different lattice parameters. After annealing, the XRD shows single-phased patterns. The patterns were well explained by the cubic FeSi-type structure. In Figure 2, lattice constants of the annealed $FeSi_{1-x}Ge_x$, determined by the Rietveld method are presented as a function of the Ge concentration x.

The lattice constants increases linearly with the Ge composition x, in accordance with the Vegard's relation. The dashed line in the figure represents the linear extrapolation from the

IOP Conf. Series: Journal of Physics: Conf. Series **868** (2017) 012009 doi:10.1088/1742-6596/868/1/012009

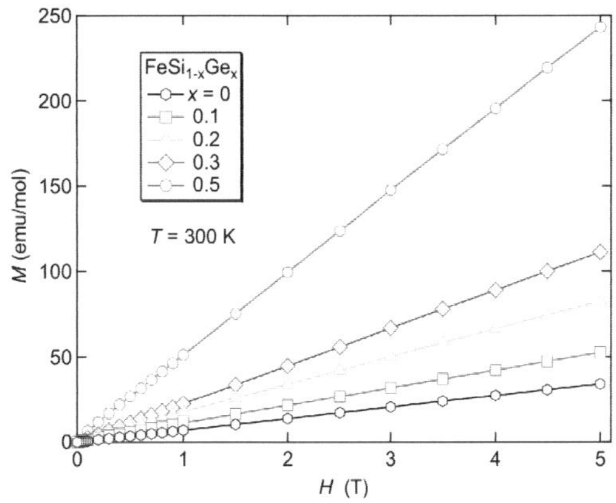

Figure 3. Magnetization M of $FeSi_{1-x}Ge_x$ annealed samples measured at T = 300 K as a function of magnetic field H.

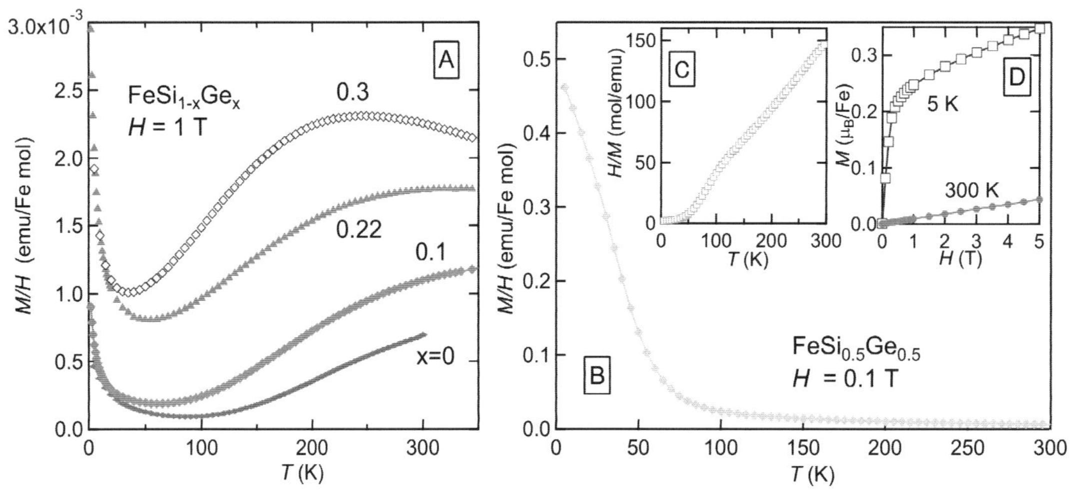

Figure 4. Temperature dependence of the magnetic susceptibility, M/H, of $FeSi_{1-x}Ge_x$. M/H of the x = 0 to 0.3 samples, shown in panel (A), are measured at H = 1 T, while that of the x = 0.5 sample, shown in panel (B), is measured at H = 0.1 T. Panel (C) shows inverse of the magnetic susceptibility, H/M, of the x = 0.5 sample as a function of temperature. Panel (D) shows magnetization of $FeSi_{0.5}Ge_{0.5}$ as a function of H.

lattice constants of FeSi, a = 4.4853 Å [8], and that of FeGe, a = 4.689 Å [22]. The lattice constants of $FeSi_{1-x}Ge_x$ obtained in the present study well agree with the linear extrapolation.

In Figure 3, magnetization M of $FeSi_{1-x}Ge_x$ annealed samples measured at T = 300 K are shown as a function of magnetic field H. M shows almost linear field dependence for all the x. This indicates these samples are free from ferromagnetic impurities such as Fe.

In Figure 4, temperature dependence of the magnetic susceptibility M/H of $FeSi_{1-x}Ge_x$ is shown.

For the samples with $x \leq 0.3$, magnetic susceptibility decreases with the decrease of T, exhibits a minimum at T = 30 to 100 K, and increases at lower temperatures. The increases at low temperature range are attributed to impurities and defects as will be suggested later. The low temperature state for $x \leq 0.3$ is therefore considered to be nonmagnetic. For the $x = 0.5$ sample, on the other hand, M/H shows a Curie-Weiss behavior from 300 K down to about 100 K, and exhibits a rapid increase below 50 K, indicating a ferromagnetic transition. The Curie temperature is roughly estimated to be T_C = 50 K from the inverse of the susceptibility H/M, shown in Figure 4 (C). In Figure 4 (D), field dependence of the magnetization M of $FeSi_{0.5}Ge_{0.5}$ is plotted. At T = 5 K, ferromagnetic behavior is observed. The spontaneous magnetic moment is estimated to be as small as 0.2 μ_B. This value is close to that obtained by Yeo et al. for x = 0.53 sample, though they reported a higher Curie temperature, T_C = 200 K [20].

Thus, the phase transition from a nonmagnetic semiconductor ($x = 0$) to a ferromagnetic metal ($x > 0.3$) reported by Yeo et al. [20] has been basically reproduced in our experiment. However, Yeo et al. reported that the phase boundary is at around $x = 0.25$ and the sample with $x = 0.27$ is a ferromagnet, while our results suggest that the phase boundary is at $x > 0.3$. This may be because our samples are polycrystalline samples, which possibly contain some distribution of Si/Ge concentration over the samples, and also because some amount of Ge may have been lost due to the evaporation during the arc melting process. In addition, Yeo et al. used the energy dispersive spectroscopy (EDS) method for the determination of the chemical compositions, and they pointed in their paper that the Si/Ge ratios can have errors of about 5 % [20].

For the sample with $x = 0.5$, M/H data above 250 K have been fit with a Curie-Weiss law, $M/H = C/(T - \theta)$ + const., where C and θ are the Curie constant and Weiss temperature, respectively. We obtained the value C = 1.86 emu K/Fe-mol, which corresponds to the effective magnetic moment of $p_{\rm eff} = 3.87\mu_B$. This value is close to that reported by Yeo et al. [20], about 3 μ_B, obtained using a single crystalline sample. When the Landé g-factor of $g = 2$ is assumed, this value of the moment close to the theoretical value $3.87\mu_B$ for the spin $S = 3/2$. For the case of itinerant electron picture, this does not necessarily indicate that each Fe ion has a local spin of $S = 3/2$. However, the result undoubtedly demonstrates that all the Fe sites carry magnetic moments, and $FeSi_{0.5}Ge_{0.5}$ does not have an energy gap.

Next, low-temperature upturn in the magnetic susceptibilities of $FeSi_{1-x}Ge_x$ samples with x = 0 to 0.3 have been analyzed with a Curie-Weiss function. In Figure 5, the fitting results are plotted. The Curie constants obtained by the low-temperature fittings are of the range of 0.01 emu K/mol. This is as small as about 0.5% of the Curie constant for $FeSi_{0.5}Ge_{0.5}$ at high temperature. We therefore can conclude that the upturns in the susceptibility at low temperature are due to small amount of impurities and defects. Thus, the intrinsic susceptibility can be estimated by subtracting the Curie-Weiss contribution at low temperature from the observed data. The corrected data are plotted in Figure 5 by blue solid lines. The intrinsic susceptibility of FeSi, shown in Figure 5(A), decreases with temperature, and becomes diamagnetic below 100 K. This corresponds to the nonmagnetic ground state of FeSi without any density of states (DOS) at the Fermi level, consistent with the previous results [2].

For $x > 0$, the intrinsic susceptibility shows similar behavior to FeSi: decreasing monotonically and reaching to a constant value below about 50 K. Notably, the constant value at low temperature shows positive values for $x > 0$ as is shown in Figure (B) to (D). This indicates that finite Pauli-paramagnetic susceptibility remains at $T = 0$ K, suggesting that the DOS at the Fermi level has a certain value for $x \geq 0.1$. Furthermore, the Pauli-paramagnetic susceptibility tends to be enhanced with the increase of x. This is qualitatively consistent with the specific heat results by Yeo et al [20]. They showed that the electronic specific heat coefficient γ is close

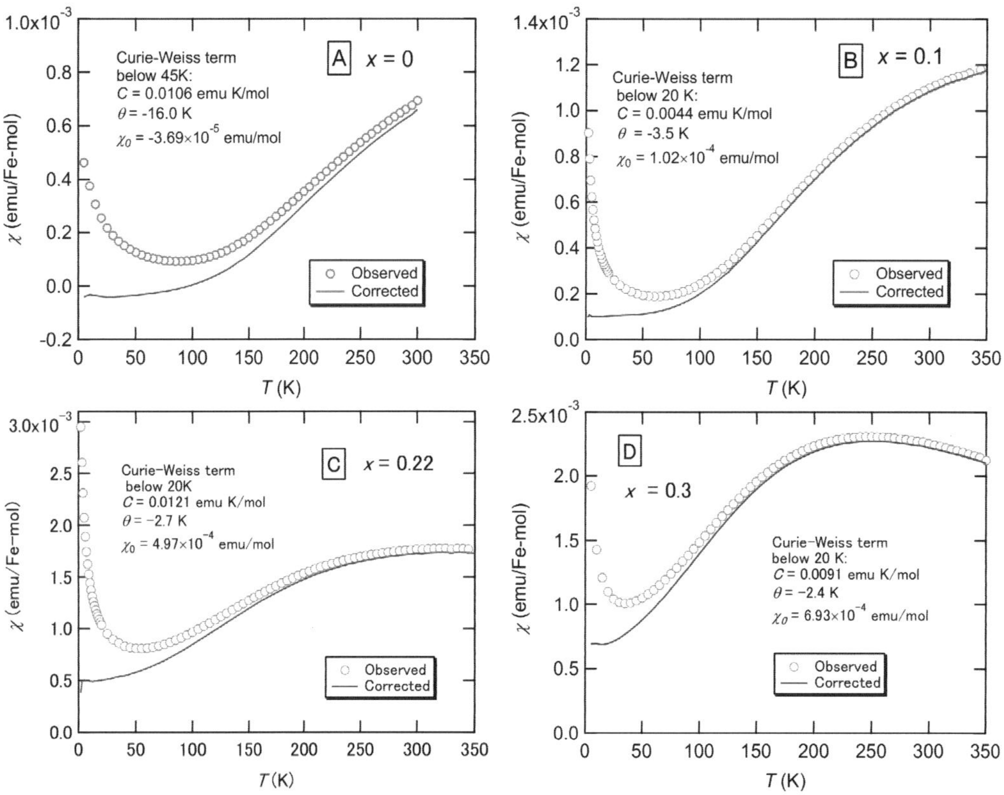

Figure 5. Magnetic susceptibility χ of $FeSi_{1-x}Ge_x$ samples with $x = 0$ to 0.3. Circles represent the observed data. Low-temperature upturn in χ has been fit with a Curie-Weiss function. Blue solid lines indicate the corrected $\chi - T$ data by subtracting the Curie-Weiss contribution at low temperature.

to zero for $x \leq 0.1$, and shows a rapid increase for $x \geq 0.2$, indicating an enhancement in the DOS at the Fermi level with the increase of x in $FeSi_{1-x}Ge_x$. In the report by Yeo et al., the γ for $x = 0.1$ is close to zero, whereas our result from the magnetic susceptibility seems to suggest that the DOS has a finite value at $x = 0.1$. This discrepancy may come from the fact that we used polycrystalline samples, where there can be distribution of Si/Ge composition and can contain much more defects that give rise to impurity states within the gap. On the other hand, the trend in γ for $x \geq 0.1$ is consistent with our magnetic susceptibility data.

Here, the low temperature susceptibility corrected after subtraction of the Curie term is attributed to the sum of the spin contribution (Pauli-paramagnetic susceptibility) and the diamagnetic contribution due to inner-core electrons. The diamagnetic susceptibility of $FeSi_{1-x}Ge_x$ was calculated using the values reported in the literature [23]. We subtracted these contribution from the χ_0 values shown inside Figure 5 to estimate the Pauli-paramagnetic susceptibility at 0 K, $\chi_{\text{spin}}(0)$. The values are plotted in Figure 6 as a function of x.

The values of $\chi_{\text{spin}}(0)$ shows a rapid increase around $x = 0.2$, and the value for $x = 0.3$ reaches relatively a large value of $\chi_{\text{spin}}(0) \simeq 0.75 \times 10^{-3}$ emu/mol. It is notable that this value of the Pauli-paramagnetic susceptibility is comparable to those of the exchange-enhanced Pauli paramagnetic systems such as YCo_2 ($\chi(0) = 1 \times 10^{-3}$ emu/Co-mol) [24] and Pd

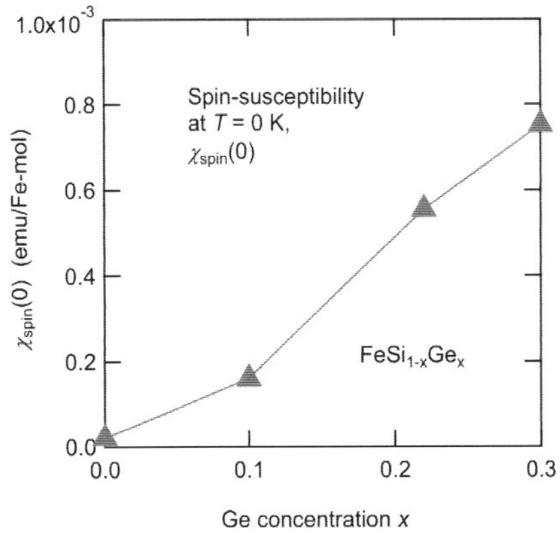

Figure 6. Spin contribution to the Pauli-paramagnetic susceptibility at $T = 0$ K, $\chi_{\rm spin}(0)$, as a function of the Ge concentration x for $FeSi_{1-x}Ge_x$.

($\chi(0) = 1.2 \times 10^{-3}$ emu/mol) [25]. These systems are well known as nearly-ferromagnetic metals. Indeed, a slight substitution of Al for Co in YCo_2 leads to an itinerant ferromagnetic phase for $x \geq 0.13$ [26]. Therefore, the transition in $FeSi_{1-x}Ge_x$ from $x = 0.3$ to 0.5 appears to be very similar to that in $Y(Co_{1-x}Al_x)_2$; from a nearly ferromagnetic metal with exchange-enhanced Pauli paramagnetic state to a weak itinerant ferromagnetic metal with a low T_C and a small spontaneous ferromagnetic moment.

In order to discuss the evolution of the magnetic correlation with the increase of Ge composition, we show the Arrot plot of $FeSi_{1-x}Ge_x$ at $T = 300$ K. This is a plot of M^2 as a function of H/M, and is written as:

$$M^2 = \zeta \frac{H}{M} - \frac{\zeta}{\chi}, \tag{1}$$

where ζ is the slop of the M^2 vs. H/M, and χ is the magnetic susceptibility. This relation is derived from the Landau expansion of the free energy F in magnetization M,

$$F(M) = \frac{1}{2\chi} M^2 + \frac{1}{4\zeta} M^4 - HM, \tag{2}$$

where the last term represents the Zeeman energy. By applying the thermal equilibrium relation $\frac{\partial F}{\partial M} = 0$, equation (1) is obtained.

The Arrot plot for $FeSi_{1-x}Ge_x$ measured at $T = 300$ K is presented in Figure 7 (A). The fitting using equation (1) has been done for the highest field regions since the linear range is very narrow because of the contribution from paramagnetic impurities. The value of the slope for $x = 0$ is $\zeta = 129.0$ emu^3/mol^{3}Oe. In the previous paper by Koyama et al. [8], they used the parameter γ as the forth order coefficient in the Landau expansion of the free energy. Using their definition, our value of ζ corresponds to $\gamma = 4.6 \times 10^3$ (Oe g^3/emu^3). This value is in good agreement with that they obtained at $T = 300$ K using an FeSi single crystalline sample. In Figure 7 (A), it is clearly seen that the slope of the Arrot plot becomes steeper with the increase of x. The value of ζ is plotted as a function of x in Figure 7 (B). ζ shows a rapid increase at $x = 0.3$, where the system is considered to be a nearly ferromagnetic state. For $x = 0.5$ where weak ferromagnetic behavior is observed, ζ is further increased but tends to be saturated. This behavior is very similar to that observed for $Y(Co_{1-x}Al_x)_2$, where ζ increases more than 10 times larger when the system approaches from nearly ferromagnetic to weak ferromagnetic state [26].

IOP Conf. Series: Journal of Physics: Conf. Series **868** (2017) 012009 doi:10.1088/1742-6596/868/1/012009

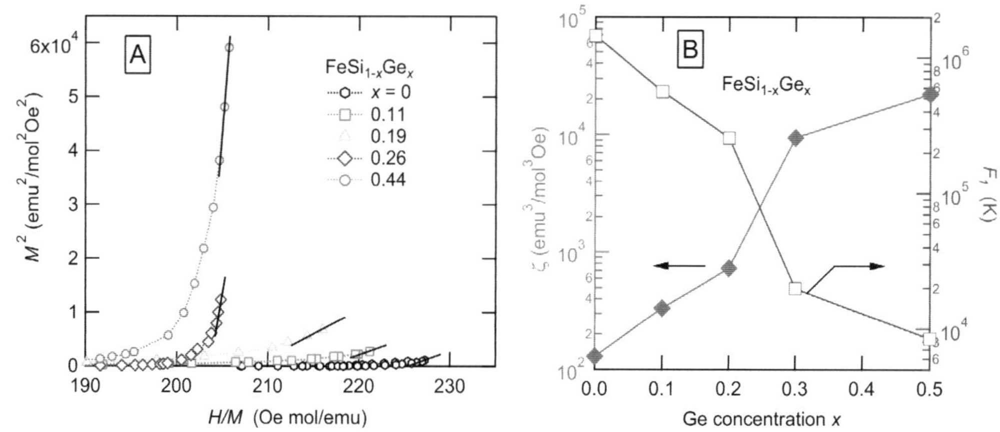

Figure 7. (A) M^2 vs. H/M plot (Arrot plot) of $\text{FeSi}_{1-x}\text{Ge}_x$ annealed samples. The magnetization data are obtained at T = 300 K. The data are shifted along the horizontal axis for clarity. (B) Values of the parameter ζ (left axis) and of F_1 (right axis) as functions of Ge concentration x. See text for details.

Next, we discuss about the x dependence of F_1, which is an important parameter in the SCR theory. We used the following relation to calculate F_1; $F_1 = 1.872 \times 10^8 \zeta^{-1}$ [27]. The values of F_1 thus estimated are plotted in Figure 7 (B). F_1 decreases drastically from 10^6 K range for x = 0 to 8.4×10^3 K for x = 0.5. The parameter F_1 is approximated to

$$F_1 = 4k_B T_A^2 / 15T_0 \tag{3}$$

in the SCR theory, where T_A and T_0 are parameters related to the spin excitation spectrum in the momentum and in the frequency space, respectively [27, 28]. When a system approaches a magnetically ordered phase from a nonmagnetic side, the spin fluctuation spectrum develops and the spectrum shape becomes quite sharp in the momentum space. This corresponds to a rapid decrease in T_A. Therefore, the rapid decrease of F_1 values shown in Figure 7 (B) is consistent with the evolution of ferromagnetic correlation. In the case of $\text{Y(Co}_{1-x}\text{Al}_x)_2$, the F_1 values show very similar behavior, decreasing from $F_1 = 4.8 \times 10^4$ K for x = 0.11, which is close to the ferromagnetically ordered phase, to $F_1 = 9.5 \times 10^3$ K for x = 0.16, where an itinerant ferromagnetic state emerges [28]. This trend of F_1 is in a very good agreement with our result shown in Figure 7 (B).

Here, we should mention that such discussion of the composition dependence of ζ and F_1 should be done using low temperature data. In our case, however, the Arrot plot using the data at T = 5 K did not show any linear range because of the dominant contribution from paramagnetic impurities. Nevertheless, the drastic decrease of F_1 in Figure 7 (B) strongly supports that the variation in magnetism of $\text{FeSi}_{1-x}\text{Ge}_x$ is understood as a transition from the nearly ferromagnetic ($x \leq 0.3$) to itinerant weak ferromagnetic metal ($x > 0.3$). To make a more accurate discussion, it would be helpful to measure the magnetization using high quality single-crystalline samples up to higher magnetic fields.

4. Conclusion

We have investigated the magnetization of $\text{FeSi}_{1-x}\text{Ge}_x$ using polycrystalline samples. A systematic evolution from a nonmagnetic semiconductor to a ferromagnetic metal with increasing

x has been confirmed, basically consistent with the previous report using single crystalline samples [20]. Our results show that the low-temperature susceptibility contains finite values of Pauli-paramagnetic susceptibility, and this contribution is increasingly enhanced with the increase of x, indicating the enhancement in the DOS at the Fermi level. For the sample with $x = 0.3$, magnetic susceptibility shows a broad maximum at around 200 K and the low temperature susceptibility is described by a large Pauli-paramagnetic susceptibility $\chi(0) = 0.75 \times 10^{-3}$ emu/mol. This behavior is very similar to that of the nearly-ferromagnetic metals such as YCo_2 and Pd. The $FeSi_{0.5}Ge_{0.5}$ sample exhibits a ferromagnetic transition at $T_C \sim 50$ K with a small spontaneous magnetic moment of 0.2 μ_B/Fe. This indicates that $FeSi_{0.5}Ge_{0.5}$ is classified as a weak itinerant ferromagnetic metal.

The evolution of ferromagnetic state with x has been analyzed using the Arrot plot (M^2 vs. H/M plot). The slope of the Arrot plot, ζ, shows a rapid increase with increasing x accross the nonmagnetic-ferromagnetic phase boundary. Using the plot, the values of F_1 have been derived, which is an important parameter in the SCR theory to describe the spin fluctuation spectrum. F_1 of $FeSi_{1-x}Ge_x$ is found to show a drastic decrease from 10^6 K for $x = 0$ to 8.4×10^3 K for $x = 0.5$. This crossover is very similar to that observed in the typical itinerant electron system $Y(Co_{1-x}Al_x)_2$ with in creasing x across the transition from the nearly ferromagnetic state to the weak ferromagnetic region. These results suggests that $FeSi_{1-x}Ge_x$ for $x > 0$ can be described as an itinerant electron system with enhanced ferromagnetic spin-fluctuations.

Acknowledgment

This work has been partly supported by Grand-in-Aid from the Japan Society for the Promotion of Science (JSPS), number 15K05190. The author thanks Namiko Onodera for the help of sample preparation and XRD measurements. The magnetization measurements using a SQUID magnetometer were carried out with the support of Tsukuba Magnet Laboratory, NIMS.

References

[1] Ishizuka K, Kiss T, Shimojima T, Yokoya T, Togashi T, Watanabe S, Zhang C Q, Chen CT, Onose Y, Tokura Y and Shin S 2005 *Phys. Rev. B* **72** 233202
[2] Jaccarino V, Wertheim G K, Wernick J H, Walker L R and Arajs S 1967 *Phys. Rev.* **160** 476
[3] Moriya T 1985 *Spin Fluctuations in Itinerant Electron Magnetism* (Berlin: Springer)
[4] Moriya T and Takahashi Y 1978 *J. Phys. Soc. Japan* **45** 397
[5] Takahashi Y and Moriya T 1979 *J. Phys. Soc. Japan* **46** 1451
[6] Shirane G, Fischer J E, Endoh Y and Tajima K 1987 *Phys. Rev. Lett.* **59** 351
[7] Tajima K, Endoh Y, Fischer J E and Shirane G 1988 *Phys. Rev. B* **38** 6954
[8] Koyama K, Goto T, Kanomata T, Note R and Takahashi Y 2000 *J. Phys. Soc. Japan* **69** 219
[9] Koyama T, Nakamura H, Kohara T and Takahashi Y 2010 *J. Phys. Soc. Japan* **79** 093704
[10] Takahashi Y 1997 *J. Phys.: Condens. Matter* **9** 2593
[11] Takahashi Y 1998 *J. Phys.: Condens. Matter* **10** L671
[12] Aeppli G and Fisk Z 1995 *Comments Condens. Matter Phys.* **16** 155
[13] Mandrus D, Sarrao J L, Migliori A, Thompson J D and Fisk Z 1995 *Phys. Rev. B* **51** 4763
[14] Schlesinger Z, Fisk Z, Zhang H T, Maple M B, DiTusa J F and Aeppli G 1993 *Phys. Rev. Lett.* **71** 1748
[15] Tsujii N, Yamaoka H, Oohashi H, Jarrige I, Nomoto D, Takahiro K, Ozaki K and Kawatsura K 2008 *Physica B* **403** 922
[16] Yamaoka H, Tsujii N, Oohashi H, Nomoto D, Jarrige I, Takahiro K, Ozaki K, Kawatsura K and Takahashi Y 2008 *Phys. Rev. B* **77** 115201
[17] Bauer E, Galatanu A, Hauser R, Reichl Ch, Wiesinger G, Zaussinger G, Galli M and Marabelli F 1998 *J. Mag. Mag. Mater.* **177-181** 1401
[18] Reichl Ch, Wiesinger G, Zaussinger G, Bauer E, Galli M and Marabelli F 1999 *Physica B* **259-261** 866
[19] Anisimov V I, Hlubina R, Korotin M A, Mazurenko V V, Rice T M, Shorikov A O and Sigrist M 2002 *Phys. Rev. Lett.* **89** 257203
[20] Yeo S, Nakatsuji S, Bianchi A D, Schlottmann P, Fisk Z, Balicas L, Stampe P A and Kennedy R J 2003 *Phys. Rev. Lett.* **91** 046401
[21] Rodrigues-Carvajal J 1993 *Physica B* **192** 55

[22] Lebech B, Bernhard J and Freltoft T 1989 *J. Phys.: Condens. Matter* **1** 6105
[23] Mendelsohn L B, Biggs F and Mann J B 1970 *Phys. Rev. A* **2** 1130
[24] Yoshimura K, Shimizu T, Takigawa M, Yasuoka H and Nakamura Y 1984 *J. Phys. Soc. Japan* **53** 503
[25] Takigawa M and Yasuoka H 1982 *J. Phys. Soc. Japan* **51** 787
[26] Yoshimura K and Nakamura Y 1985 *Solid State Commun.* **56** 767
[27] Takahashi Y and Yoshimura K 2012 *Itinerant electron magnetism and spin fluctuation* (in Japanese, Uchida Rokakuho Publishing Co., Tokyo)
[28] Yoshimura K, Takigawa M, Takahashi Y, Yasuoka H and Nakamura Y 1987 *J. Phys. Soc. Japan* **56** 1138

Study of itinerant electronic ferromagnetism in layered compounds Ae_2ScO_3CoPn (Ae = Sr, Ba, Pn = P, As)

Hiroto Ohta, Shohei Ido, Shuku Hashizume, and Hiroko Aruga Katori

Institute of Engineering, Division of Advanced Applied Physics, Tokyo University of Agriculture and Technology, Nakacho 2-24-16, Koganei, Tokyo 184-8588, Japan

E-mail: h-ohta@cc.tuat.ac.jp

Abstract. We successfully synthesized polycrystalline samples of new layered compounds Ae_2ScO_3CoP (Ae = Sr, Ba) and Ba_2ScO_3CoAs, members with CoPn conduction layers, and measured magnetizations of them. As results, magnetic parameters, i.e. the Curie temperature and the Weiss temperature, are found not to be correlative with lattice parameters a nor c in the case of Ae_2ScO_3CoPn compounds, including Sr_2ScO_3CoAs, unlike the case of three dimensional itinerant electronic ferromagnets. We suggested that structural parameters of $CoPn_4$ tetrahedron are good parameters to relate to magnetic parameters in the case of compounds with CoPn layers.

1. Introduction

Itinerant electronic magnetism has been one of important topics not only in the eld of magnetism but also in the eld of strongly correlated electronic systems. Theory of itinerant electronic magnetism was developed by T. Moriya and his co-workers and the self-consistent renormalization (SCR) theory for spin fluctuations was established for itinerant electronic ferromagnetism (IEF). [1] The theory for IEF has been still developed by Y. Takahashi [2] and several important problems which Moriya's theory have, e.g. critical phenomenon around the Curie temperature (T_C), magnetocaloric e ect and so on, have been solved in Takahashi's theory. [3] In the case of isotropic band structure, the understanding of IEF has grown from the viewpoint of theory and experiment while in the case of anisotropic band structure the understanding of IEF is not enough since experimental researches have not been done systematically.

Recently, layered compounds LaCoPnO (Pn = P or As), which have ZrCuSiAs-type crystal structure, were found to show IEF with T_C being 30 K (for Pn = P) or 55 K (for Pn = As). [4] Since $3d$ electrons of Co, which delocalize within CoPn layers, are responsible for magnetism in these compounds, LaCoPnO have been thought as quasi two dimensional IEF compounds with anisotropic band structures. Spin fluctuations of LaCoPnO were reported to actually have two dimensional nature in the paramagnetic state. [5, 6] Sr_2ScO_3CoAs, which has Sr_2GaO_3CuS-type crystal structure [7, 8], has CoAs conductive layers as well as LaCoAsO but the distance between CoAs layers is about twice larger than that of LaCoAsO. [9] Although this compound does show ferromagnetic ordering at T_C = 48 K, there was reported that two dimensionality of itinerant electrons in CoAs layers enhances in Sr_2ScO_3CoAs in comparison with LaCoAsO. [10] Among compounds with Co*Pn* layers, those with Sr_2GaO_3CuS-type crystal structure have the most largest distance between Co*Pn* layers for now, so we think such compounds are put on a good stage for studying IEF in anisotropic band structures. In addition, since compounds with Sr_2GaO_3CuS-type structure are composed of 5 elements, there seems to have the variety as the compounds which contain just Co as a magnetic element. For these reasons, we studied layered compounds Ae_2ScO_3CoPn (Ae = Sr, Ba, Pn = P, As) to clarify nature of itinerant electronic magnetism in anisotropic structures.

In this paper, we report results of synthesis and magnetic measurements of new layered compounds Ba_2ScO_3CoAs, Ae_2ScO_3CoP (Ae = Sr or Ba). We also discuss relation between itinerant electronic ferromagnetism and crystal structure of layered compounds with Co*Pn* (Pn = P or As) conduction layers.

2. Experiments

Polycrystalline samples of Ba_2ScO_3CoAs, Ae_2ScO_3CoP (Ae = Sr or Ba) were synthesized by a conventional solid state reaction method. First, powders of CoAs and CoP were synthesized by mixing powders of Co (purity: 99.9 %) and As (99.9 %) or P (99.9999 %) in the ratio of 1 : 1 and ring them at 800 °C for 12 hs in evacuated silica tubes. Then, obtained Co*Pn* powders were mixed with *Ae* (99 %), *Ae*O (99 %), and Sc_2O_3 (99.9 %) in the appropriate ratio and these mixtures were red at 1,200 °C for 12 hs in evacuated silica tubes. This step was repeated once or twice to achieve good homogeneity. Samples were handled in a glove box lled with highly pure N_2 gas. Obtained samples were characterized by measuring powder X-ray di raction (XRD) with Cu K_α radiation. Magnetization of samples were measured by using vibrating sample magnetometer ($MagLab^{\mathrm{VSM}}$, Oxford Instruments).

3. Results and Discussion

Figure 1 (a) shows the results of powder XRD measurements of the samples. As shown in the gure, we successfully synthesized each compound though Sr_2ScO_3CoP and Ba_2ScO_3CoP contain small amount of $SrCo_2P_2$ and $BaCo_2P_2$, respectively, as impurities. Since these impurities are paramagnets, [11, 12] their contribution to magnetization of each sample is thought to be negligible. From the data of XRD, we estimated lattice parameters along a and c-axes of each compound. The estimated values are listed in Table 1. For the case of Sr_2ScO_3CoP, our result is consistent with that reported by Okada *et al.* [13] We also shows the values of a and c for four compounds, i.e. Ba_2ScO_3CoAs, Sr_2ScO_3CoAs, Ba_2ScO_3CoP and Sr_2ScO_3CoP, in Fig. 1 (b). Both the parameters obviously depend primarily on ion size of alkali earth metal elements and secondly on that of pnictogen elements. By changing alkali earth metal elements from Sr to Ba and pnictogen elements from P to As, the values of a and c change about 4 % and 9 %, respectively. Such a large change in size of unit cell promises variation of magnetism in these compounds, since IEF strongly correlates to cell volume of the system in usual.

Figure 2 (a) shows temperature (T) dependence of magnetization (M) of Ba_2ScO_3CoAs measured at H = 1 T together with that of Sr_2ScO_3CoAs (denoted by Sr/As in the gure) for comparison. Clear ferromagnetic ordering was observed below about 40 K. Figure 2 (b) shows isothermal magnetization curves of Ba_2ScO_3CoAs at various temperatures. Below 40 K, spontaneous magnetizations were observed in consistency with the result of T dependence of M. The Curie temperature (T_{C}) of this compound was estimated as 32 K, which is less than that of Sr_2ScO_3CoAs (T_{C} = 48 K).

Figure 3 (a) shows T dependence of M of Sr_2ScO_3CoP measured at H = 1 T. In the low temperature region, M of the sample increases with decrease of T and seems to show ferromagnetic ordering below about 10 K. Figure 3 (b) shows isothermal magnetization curves

IOP Conf. Series: Journal of Physics: Conf. Series **868** (2017) 012010 doi:10.1088/1742-6596/868/1/012010

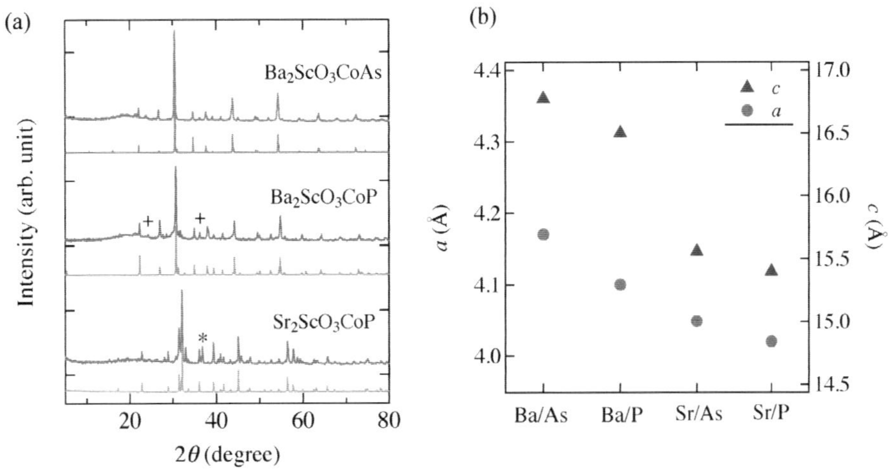

Figure 1. (a) Results of powder XRD measurements of the samples. For each sample, upper and lower row show observed and calculated data, respectively. Peaks indicated by "+" and "*" correspond to the patterns of $BaCo_2P_2$ and $SrCo_2P_2$, respectively. (b) Lattice parameters along a and c-axes for Ae_2ScO_3CoPn (abbreviated as Ae/Pn).

Table 1. Estimated lattice parameters from XRD measurements. V is cell volume.

	a (Å)	c (Å)	V (Å^3)
Ba_2ScO_3CoAs	4.170	16.773	291.7
Sr_2ScO_3CoAs [10]	4.049	15.557	255.0
Ba_2ScO_3CoP	4.099	16.504	277.3
Sr_2ScO_3CoP	4.020	15.412	249.1

at various temperatures. At $T = 2$ K, magnetic curve shows ferromagnetic enhancement as if the system is in ferromagnetic ordered state as the ground state. However, spontaneous magnetization was not observed at this temperature.

Figure 4 (a) shows T dependence of M of Ba_2ScO_3CoP measured at $H = 2$ T. In contrast with other samples, this sample does not show ferromagnetic ordering at all down to 2 K. Figure 4 (b) shows isothermal magnetization curves at various temperatures. Although magnetization curve at $T = 2$ K is convex upward, spontaneous magnetization does not observed down to 2 K consistent with the result of T dependence of M.

Figure 5 shows temperature dependence of H/M of the three samples together with that of Sr_2ScO_3CoAs for comparison. H/M of all the compounds show linear temperature dependence, meaning that the Curie Weiss law holds. Magnetic parameters, i.e. the e ective Bohr magneton number ($P_{\rm eff}$) and the Weiss temperature (θ), were estimated by tting a function of the Curie Weiss law to the data and listed in Table 2. Almost constant value of $P_{\rm eff}$ and systematic change of θ among these compounds are well explained by Moriya's SCR theory for ferromagnetic spin fluctuations. [1] $P_{\rm eff}$ of Ba_2ScO_3CoP is smaller compared with the others. Possibly, this is because an impurity phase exists in Ba_2ScO_3CoP. In the Moriya's theory, θ has an important meaning. The absolute value of θ corresponds to the "distance" from a quantum critical point

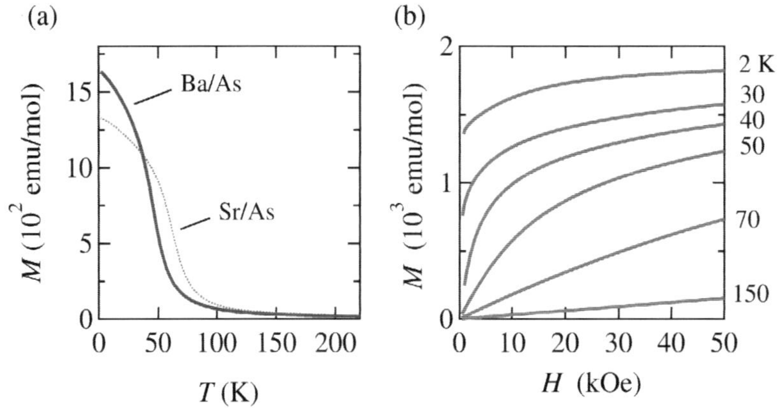

Figure 2. (a) T dependence of M of Ba_2ScO_3CoAs measured at $H = 1$ T shown together with that of Sr_2ScO_3CoAs for comparison. (b) Isothermal magnetization curves at various temperatures.

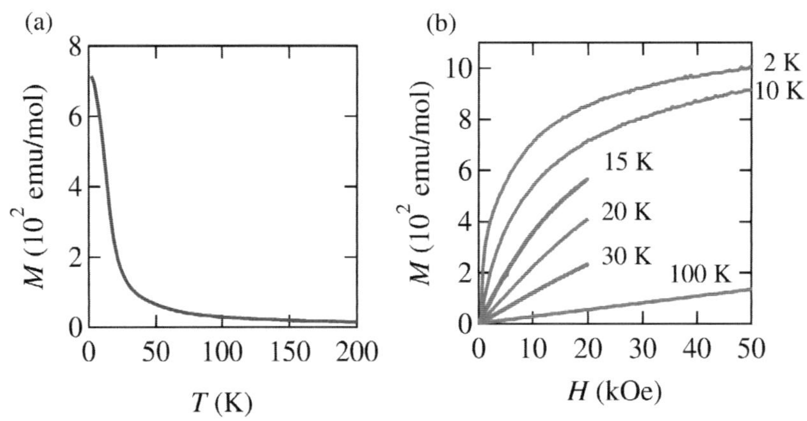

Figure 3. (a) T dependence of M of Sr_2ScO_3CoP measured at $H = 1$ T. (b) Isothermal magnetization curves at various temperatures.

(QCP) and the sign of θ shows whether the system is in ordered state or disordered state: when $\theta > 0$ the system is in ferromagnetically ordered state while when $\theta < 0$ the system is in disordered (paramagnetic) state. As shown in Figure 6, electronic states of these compounds expressed by U/W are located in the narrow region around QCP on the electronic phase diagram of IEF, and especially electronic state of Sr_2ScO_3CoP is in the vicinity of QCP of IEF.

In the case of compounds with isotropic band structure, enlargement of cell volume leads to decrease of θ or depression of ferromagnetism, since energy width of conduction band W correlates to the cell volume: the larger cell volume corresponds to the smaller W (larger U/W and θ) and vice versa. However, electronic systems of Ae_2ScO_3CoPn apparently do not have such a relationship between cell volume and θ. In principle, strength of ferromagnetism, or θ, should correlate to U/W, we think that W does not simply relate to V any more in the case of Ae_2ScO_3CoPn. Such anomalous relationship has been reported in compounds with CoPn layers, [14, 15] indicating that this anomalous relationship is due to layered crystal structure.

IOP Conf. Series: Journal of Physics: Conf. Series **868** (2017) 012010 doi:10.1088/1742-6596/868/1/012010

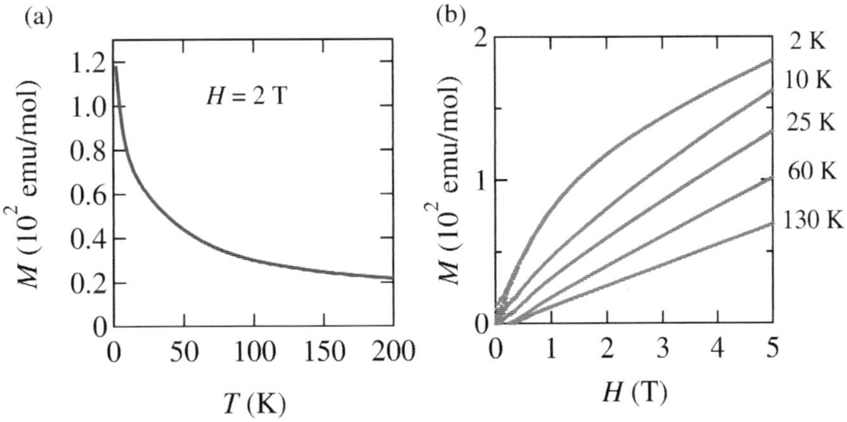

Figure 4. (a) T dependence of M of Ba_2ScO_3CoP measured at $H = 2$ T. (b) Isothermal magnetization curves at various temperatures.

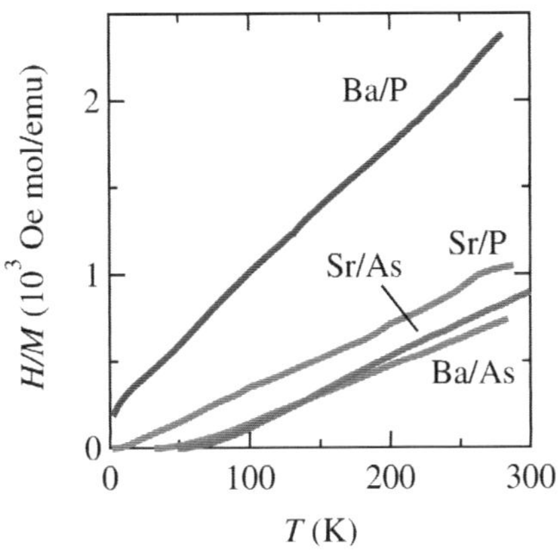

Figure 5. T dependence of H/M of the samples together with that of Sr_2ScO_3CoAs. Ae/Pn in the gure denote Ae_2ScO_3CoPn.

Table 2. Magnetic parameters of the compounds: the e ective Bohr magneton number $P_{\rm eff}$, the Weiss temperature θ and the Curie temperature $T_{\rm C}$.

	$P_{\rm eff}$	θ (K)	$T_{\rm C}$ (K)
Ba_2ScO_3CoAs	1.57	56	32
Sr_2ScO_3CoAs [10]	1.34	78	48
Ba_2ScO_3CoP	1.04	-36	-
Sr_2ScO_3CoP	1.47	6.8	0

In the case of Fe-analogue, relationship between superconducting transition temperature T_c and local structural parameters relating to $FeAs_4$ tetrahedron has been proposed. [16] In the

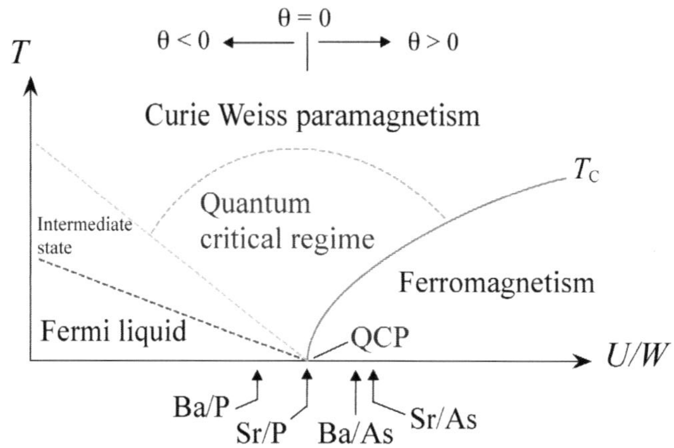

Figure 6. Schematic phase diagram of IEF around QCP. [1] Solid line shows a phase boundary while dotted lines just show cross over regions. U and W are the Coulomb repulsion energy and energy width of conduction band, respectively. Ae/Pn in the gure denote Ae_2ScO_3CoPn.

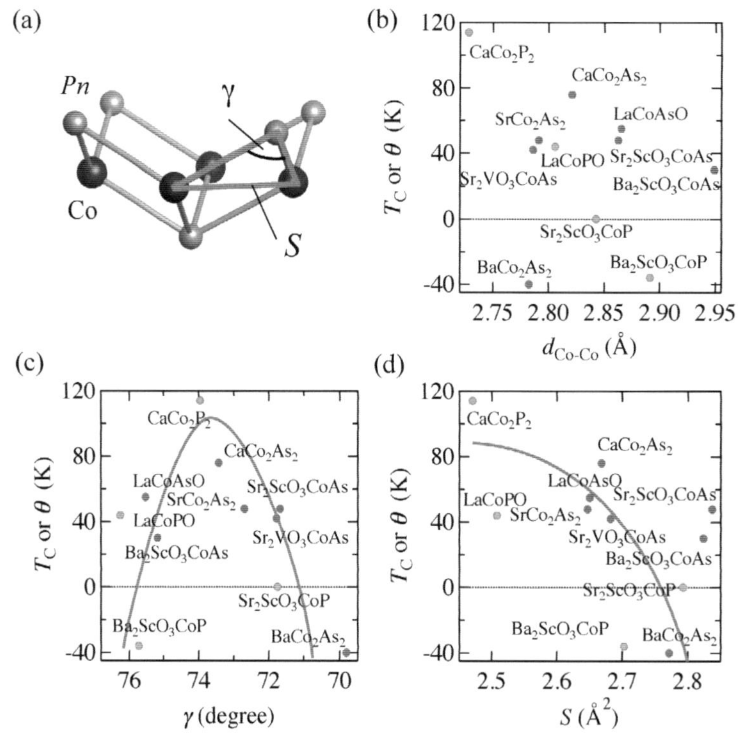

Figure 7. (a) local structure of CoPn layer. (b)-(d) T_C or θ plotted against (b) d_{Co-Co}, (c) S and (d) γ. Data are after Ref. [4, 14, 17, 18, 19, 20]. In (b)-(d), red and orange markers show the data of arsenide and phosphide compounds, respectively.

track of the Fe-analogue, we checked relationship between θ and local structural parameters relating to CoPn_4 tetrahedron. Among such parameters, we tentatively chose two parameters

S and γ, which are area of triangle of Co-Pn-Co and angle of Co-Pn-Co respectively as shown in Figure 7 (a). Figure 7 (b) shows T_C or θ of compounds with CoPn layers plotted against d_{Co-Co}, where d_{Co-Co} is the distance between nearest neighbor Co atoms within the CoPn layer and can be expressed by $d_{Co-Co} = a/\sqrt{2}$. There does not seem to be a correlation between two parameters. Figure 7 (c) and (d) show T_C or θ plotted against S or γ, respectively. As shown by solid lines in each panel, strength of magnetism seems to correlate with local structural parameters: the data of T_C or θ show a parabolic curvature against γ while they show a monotonic decrease with increase of S. These correlations can roughly be understood as that energy width of conduction band W depends on how p-orbitals of Pn are hybridized with $d_{x^2-y^2}$ orbital of Co, since overlap integral of p-orbitals of Pn and $d_{x^2-y^2}$ orbital of Co depends on both relative angle between these orbitals and distance between Pn and Co. It should be noted that this conclusion must be veri ed experimentally by using more precise magnetic and structural parameters and also theoretically by doing precise band calculation and so on.

4. Summary

We successfully synthesized polycrystalline samples of new layered compounds Ba_2ScO_3CoAs and Ae_2ScO_3CoP (Ae = Sr or Ba). We found that the Weiss temperatures of these compounds are ranging from about -40 to 60 K, indicating that these compounds are located in a narrow region around quantum critical point of itinerant electronic ferromagnetism. We proposed relationships between strength of magnetism and local structural parameters.

Acknowledgments

This work was supported by Grant-in-Aid for Young Scientists (B) from Japan Society for the Promotion of Science (Grant No. 24760534).

References

[1] Moriya T 1985 *Spin Fluctuations In Itinerant Electron Magnetism* (Springer-Verlag Berlin Heidelberg) ISBN 978-3-642-82501-9
[2] Takahashi Y 1986 *J. Phys. Soc. Jpn.* **55** 3553
[3] Takahashi Y 2013 *Spin Fluctuation Theory of Itinerant Electron Magnetism* (Springer-Verlag Berlin Heidelberg) ISBN 978-3-642-36665-9
[4] Yanagi H, Kawamura R, Kamiya T, Kamihara Y, Hirano M, Nakamura T, Osawa H and Hosono H 2008 *Phys. Rev. B* **77** 224431
[5] Sugawara H, Ishida K, Nakai Y, Yanagi H, Kamiya T, Kamihara Y, Hirano M and Hosono H 2009 *J. Phys. Soc. Jpn.* **78** 113705
[6] Ohta H and Yoshimura K 2009 *Phys. Rev. B* **79** 184407
[7] Zhu W J and Hor P H 1997 *Inorg. Chem.* **36** 3576
[8] Tegel M, Hummel F, Lackner S, Schellenberg I, P ottgen R and Johrendt D 2009 *Z. Anorg. Allg. Chem.* **635** 2242
[9] Xie Y L, Liu R H, Wu T, Wu G, Song Y A, Tan D, Wang X F, Chen H, Ying J J, Yan Y J, Li Q J and Chen X H 2009 *EPL* **86** 57007
[10] Ohta H, Noguchi D, Nabetani K and Aruga Katori H 2013 *Phys. Rev. B* **88** 094441
[11] Reehuis M and Jeitschko W 1990 *J. Phys. Chem. Solids* **51** 961
[12] Imai M, Ohta H, Michioka C, Ueda H and Yoshimura K 2012 *Proceedings of the 2012 Powder Metallurgy World Congress and Exhibition* 16F-T14-121
[13] Okada S, Kamihara Y, Ohkubo N, Ban S, Matoba M and Atou T 2012 *J. Phys.: Conf. Ser.* **400** 022091
[14] Jia S, Williams A J, Stephens P W and Cava R J 2009 *Phys. Rev. B* **80** 165107
[15] Sefat A S, Singh D J, Jin R, McGuire M A, Sales B C and Mandrus D 2009 *Phys. Rev. B* **79** 024512
[16] Lee C H, Iyo A, Eisaki H, Kito H, Fernandez-Diaz M T, Ito T, Kihou K, Matsuhata H, Braden M and Yamada K 2008 *J. Phys. Soc. Jpn.* **77** 083704
[17] Reehuis M, Jeitschko W, Kotzyba G, Zimmer B and Hu X 1998 *J. Alloys Comp.* **266** 54
[18] Yu-Min Q and Gang X 2011 *Acta Phys. Sin.* **60** 067101
[19] P sterer M and Nagorsen G 1980 *Z. Naturforsch.* **35b** 703
[20] Imai M, Michioka C, Ohta H, Matsuo A, Kindo K, Ueda H and Yoshimura K 2014 *Phys. Rev. B* **90** 014407

IOP Conf. Series: Journal of Physics: Conf. Series **868** (2017) 012011 doi:10.1088/1742-6596/868/1/012011

Spin fluctuation and giant magnetoresistance in CeSi

Satoru Noguchi
Nanoscience and Nanotechnology Research Center, Osaka Prefecture University, Sakai, Osaka 599-8570, Japan

E-mail: noguchi@21c.osakafu-u.ac.jp

Abstract. We measured temperature-dependent resistivity of a CeSi single crystal under several magnetic fields in order to study the effect of the magnetic field on the spin fluctuation in the commensurately modulated antiferromagnetic state of CeSi. Resistivity under each magnetic field exhibits a clear T^2 dependence at low temperatures. The coefficient A of T^2 term decreases monotonously with increasing magnetic field. The residual resistivity at 0 T is steeply reduced around 1 T, and almost constant above 1 T. Temperature dependence of the magnetoresistivity near T_N is compared with that of the isothermal change in the magnetic entropy.

1. Introduction

CeSi crystallizes in the orthorhombic FeB type structure [1] and it exhibits an antiferromagnetic order below T_N =6 K [2,3]. The antiferromagnetic ordering is a commensurate sinusoidal or square wave magnetic structure characterized by the wave vector $\boldsymbol{q}$ = (0, 1/2, 1/16) [4,5], showing a metamagnetic transition at B_c = 0.6 T [2,3]. A large anisotropy has been observed by our group in the magnetization, magnetic susceptibility and resistivity measurements [6]. A large negative magnetoresistivity (MR) attained to $-\Delta\rho/\rho(0) \sim 0.7$ was observed for the field direction parallel to the b axis. It shows the M^4 dependence at 4.2 K, where M is the magnetization of CeSi for $B//b$ [7]. A peculiar field dependence of the anomalous Hall resistivity has been found at 4.2 K, changing the sign at 5 T [8]. Temperature dependence of the Hall coefficient fits to the susceptibility data above 30 K for the b-axis with a proper scale factor and offset [8], suggesting the strong exchange interaction between the 4f spins and the conduction electrons. Kondo effect is very weak in CeSi [3]. The valence-band electronic structures have been investigated using ultraviolet photoemission spectroscopy, which shows that the 4f electrons hybridize with the valence electrons though Kondo resonance peak was not detected [9]. They found from temperature dependence of high-resolution resonant photoemission spectra of CeSi that the heavy fermion behaviour is strongly suppressed by the antiferromagnetic ordering due to the RKKY interaction [10]. Recently, magnetocaloric effect (MCE: the isothermal change in magnetic entropy) was studied in the CeSi polycrystals by Snyman *et al.* [11]. They calculated the anisotropic exchange parameters from the analysis of the MCE data. More recently, the MCE data calculated from the magnetization measurements were reported by Wang *et al.* [12].

In this paper, we present temperature-dependent resistivity of the CeSi single crystal under several magnetic fields in order to study the effect of the magnetic field on the spin fluctuation in the commensurately modulated antiferromagnetic state of CeSi. We note that temperature dependence of the MR around T_N is similar to that of the MCE, which suggests that the RKKY interaction is dominant in this compound.

2. Experimental

The CeSi single crystals were prepared by the Czochralski pulling method in a tri-arc furnace. The resistance measurements were performed by a standard dc 4-terminal method in a conventional superconducting magnet system up to 8 T. The magnetocaloric effect was obtained from the data of the specific heat measurements performed previously by our group [8].

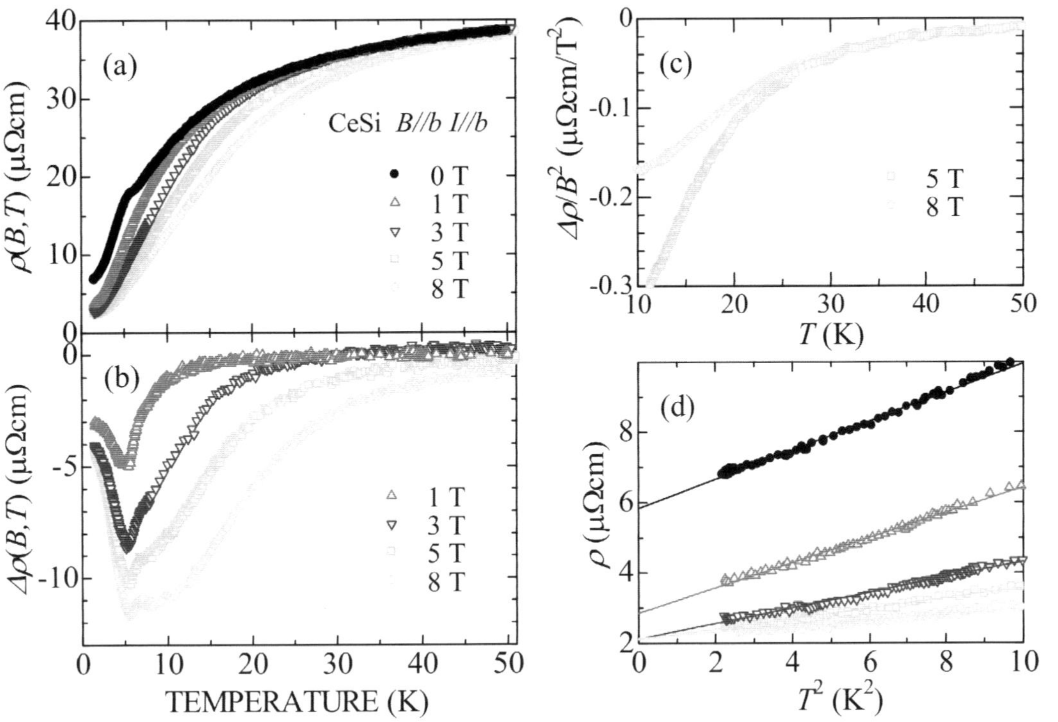

Figure 1. (a): Temperature-dependent resistivity of CeSi single crystal under magnetic fields along the *b*-axis. (b): Temperature-dependent magnetoresistivity obtained by subtracting the zero-field data. (c): Temperature-dependent magnetoresistivity divided to B^2 for B = 5 T and 8 T. Two data are almost the same values above 25 K. (d): The detail of temperature-dependent resistivity at low temperatures as a function of T^2.

3. Results and discussions

Temperature-dependent resistivity of the CeSi single crystal under the magnetic fields B = 0, 1, 3, 5 and 8 T along the *b*-axis was measured below 50 K down to 1.5 K, as shown in Fig. 1 (a). The current parallel to the magnetic field was applied so as to avoid mixing the orbital MR. The data at zero magnetic field shows a small hump at 6 K, corresponding to the antiferromagnetic ordering temperature T_N, which suggests the change in density of states with the magnetic ordering. By subtracting the zero-field data from the data under each magnetic field, resistivity due to the phonon scattering and the nonmagnetic impurity scattering are cancelled, and subsequently temperature dependence of the MR is extracted as shown in Fig. 1 (b). A large negative MR appears in the temperature below 50 K. Broad peaks observed around 10 K are attributed to the effect of the magnetic field that suppresses the fluctuation of the paramagnetic spins. A small peak at 6 K is caused by the antiferromagnetic transition observed as the small hump in the zero-field data. Fig. 1 (c) shows the MR divided by B^2 for 5 T and 8 T. Both curves are in good accordance at high temperatures above 25 K, where the magnetization M is proportional to the applied field. This suggests that the MR is proportional to M^2 in the temperature range. Below 25 K, both curves are not the same trace because

M deviates from the proportional relation to B in the Brillouin function at low temperatures under high magnetic fields. The M^2 dependence of the MR is commonly recognized in the compounds showing a large negative MR such as the GMR [13] and the CMR [14] compounds. At 4.2 K, we previously reported a large negative magnetoresistance proportional to M^4 [7]. So, the MR dependence of M^2 changes to that of M^4 between 4.2 K and 20 K at most.

The data at low temperatures below 3 K obey clear T^2 dependence as shown in Fig. 1 (d). By fitting the simple equation,

$$\rho = \rho_0 + AT^2 \tag{1}$$

we obtained both the coefficient of the T^2 term A and the residual resistivity ρ_0 as a function of the magnetic field as shown in Fig. 2. The value of ρ_0 at 0 T is almost 3 times larger than the values above 1 T. This change is attributed to the metamagnetic transition from the commensurately modulated antiferromaget to the field-induced paramagnet in CeSi. This suggests that the change in density of states associated with the antiferromagnetic ordering is reset by the magnetic field. Above 1 T, the value of ρ_0 is almost constant. A shows a monotoous decrease with increasing magnetic field, which may be attributed to the suppression of the spin fluctuation by applying magnetic field in the field-induced paramagnetic state. Field dependence of A can be qualitatively explained by the self-consistent renormalization theory of spin fluctuations [15].

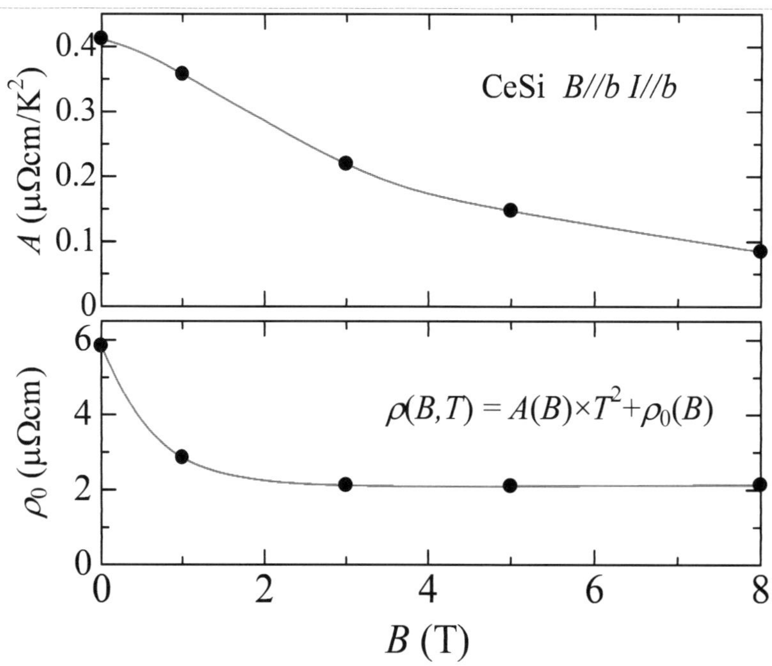

Figure 2. Magnetic field dependence of the parameters A and ρ_0. Solid curves are drawn by a spline interpolation.

We previously measured the iso-field specific heat of the CeSi single crystal for $B//b$-axis, where the lattice contribution was subtracted by using the specific heat of LaSi [8]. From the data, we calculated temperature dependence of the magnetic entropy S_{mag} as shown in Fig. 3(a). The value of the magnetic entropy under zero magnetic field at T_N is almost corresponding to $R\ln 2$, where R is the gas constant. This suggests a doublet ground state of 4f localized electron in CeSi as previously pointed out [6]. By subtracting the zero-field data from the data under each magnetic field, temperature dependence of the MCE ΔS_{mag} is extracted as shown in Fig. 3 (b). A small positive peak below T_N for 1 T and large negative peaks around T_N appear, which is well in agreement with the MCE data of the CeSi polycrystal by Snyman *et al.* [11], although the absolute value of ΔS_{mag} is almost twice larger than their data. The absolute value of the MCE for 5 T is -2.40 J/K/mol at 7 K, which is in good agreement with the MCE data of -13.7 J/K/kg for 5 T evaluated from the isothermal magnetization measurements of CeSi by Wang *et al.* [12]. These suggest that the MCE is anisotropic.

The temperature dependence of the MCE data are well corresponding to that of the MR in Fig. 1 (b). Campoy *et al.* have showed that MR is a probe to the MCE in ferromagnetic rare-earth aluminum compounds [16]. In the antiferromagnetic case, the scaling between the MR and the MCE may not be so simple because the metamagnetic or spin-flop transition generally occurs in the antiferromagnets. Theoretical studies are required.

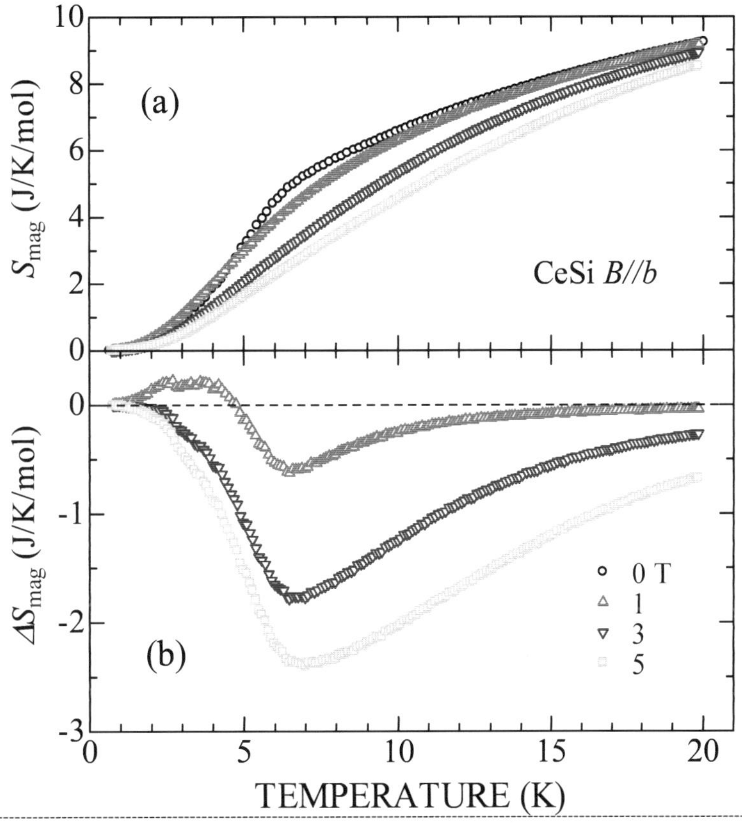

Figure 3. (a): Temperature-dependence of the magnetic entropy of the CeSi single crystal under magnetic fields along the *b*-axis. (b): Temperature-dependent MCE obtained by subtracting the zero-field data.

4. Conclusion

We measured temperature-dependent resistivity of a CeSi single crystal under several magnetic fields in order to study the effect of the magnetic field on the spin fluctuation in the commensurately modulated antiferromagnetic state of CeSi. Resistivity under each magnetic field exhibits a clear T^2

dependence at low temperatures. The coefficient A of T^2 term decreases monotonously with increasing magnetic field. This is qualitatively explained by the self-consistent renormalization theory of spin fluctuations. The residual resistivity at 0 T is steeply reduced around 1 T, and almost constant above 1 T. This suggests that the change in density of states associated with the antiferromagnetic ordering is reset by the magnetic field. Temperature dependence of the magnetoresistivity near T_N is compared with that of the isothermal change in the magnetic entropy, suggesting the strong exchange interaction between the 4*f* spins and the conduction electrons.

5. References

[1] Hohnke D H and Parthe E 1966 *Acta Crystallogr.* **20** 572.
[2] Shaheen S A 1987 *Phys. Rev.* B **36** 5472.
[3] Shaheen S A 1988 *J. Appl. Phys.* **63** 3411.
[4] Schobinger-Papamantellos P and Buschow K H J 1994 *J. Magn. Magn. Mater.* **130** 242.
[5] Schobinger-Papamantellos P, Kenzelmann M, Schenck A, Gygax F N, Buschow K H J and Ritter C 2004 *Physica* B **349** 100.
[6] Noguchi S, Okuda K, Nojiri H and Motokawa M, 1998 *J. Magn. Magn. Mater.* **177-181** 403.
[7] Noguchi S 2003 *Physica* B **329-333** 491.
[8] Noguchi S, Miyagawa S, Katori H A and Ishida T 2004 *J. Magn. Magn. Mater.* **272-276** e1533.
[9] Mimura K, Takase T, Mizohata H, Taguchi Y, Ichikawa K, Takeda Y, Arita M, Shimada K, Namatame H, Noguchi S, Okuda K, Taniguchi M and Aita O 2001 *J. Electron Spectrosc. Relat. Phenom.* **114-116** 723.
[10] Mimura K, Noguchi S, Suzuki M, Higashiguchi M, Shimada K, Ichikawa K, Taguchi Y, Namatame H, Taniguchi M and Aita O 2005 *J. Electron Spectrosc. Relat. Phenom.* **144-147** 715
[11] Snyman J L, Carleschi E, Doyle B P and Strydom A M 2013 *J. Appl. Phys.* **113** 17A903.
[12] Wang L C, Dong Q Y, Lu J, Shao X P, Mo Z J, Xu Z Y, Sun J R, Hu F X and Shen B G 2014 *J. Alloys Comp.* **587** 10.
[13] Xiao J Q, Jiang J S and Chien C L 1992 *Phys. Rev. Lett.* **68** 3749.
[14] Urushibara A, Moritomo Y, Arima T, Asamitsu A, Kido G and Tokua Y 1995 *Phys. Rev.* B **51** 14103.
[15] Ueda K 1977 *J. Phys. Soc. Jpn.* **43** 1947.
[16] Campoy J C P, Plaza E J R, Coelho A A and Gama S 2006 *Phys. Rev.* B **74** 13440.

Anisotropy of Spin Fluctuations in a Tetragonal Heavy Fermion Antiferromagnet $CeRhAl_4Si_2$

H. Sakai[1], **T. Hattori**[1], **Y. Tokunaga**[1], **S. Kambe**[1], **N. J. Ghimire**[2,3], **F. Ronning**[2], **E. D. Bauer**[2], **and J. D. Thompson**[2]

[1] Advanced Science Research Center, Japan Atomic Energy Agency, Tokai, Ibaraki, 319-1195, Japan.

[2] Condensed Matter and Magnet Science, Los Alamos National Laboratory, Los Alamos, NM 87545, USA.

E-mail: sakai.hironori@jaea.go.jp

Abstract. An antiferromagnetic (AFM) Kondo lattice compound $CeRhAl_4Si_2$, which exhibits successive AFM transitions at T_{N1}=14 K and T_{N2}=9 K in zero external field, has been microscopically investigated by means of ^{27}Al nuclear magnetic resonance (NMR) technique. In the high temperature range, magnetic excitations of $4f$ electrons can be well explained by isotropic localized spin fluctuations. Below ~50 K, it begins to show a characteristic anisotropy of spin fluctuations, which suggests a competition between spin fluctuations and nesting instability in this system.

1. Introduction

The tetragonal compound $CeRhAl_4Si_2$, which is one of the series $CeT_nAl_{2n+2}Si_2$ (T=Rh, Ir, Pt), has been reported as a new antiferromagnetic (AFM) Kondo lattice with successive AFM transitions of T_{N1}=14 K and T_{N2}=9 K at zero external field. [1, 2] These $CeTAl_4Si_2$ (T=Rh, Ir, Pt) materials crystallize in the tetragonal KCu_4S_3-type structure with space group $P4/mmm$, as displayed in Fig. 1 for $CeRhAl_4Si_2$. In this quaternary compound, the square lattices of Ce-Ce with a=4.22 Å are well separated from each other along the c-axis of 8.01 Å. Due to the layered crystal structure, a quasi-two-dimensional electronic structure might be expected. Indeed, density functional theory (DFT) calculations using the generalized gradient approximation (GGA) predict a quasi-two-dimensional character of the Fermi surface with a propensity for nesting in the ab plane as well as along the c-axis. [1] From magnetization measurements on $CeRhAl_4Si_2$, the effective moment in the paramagnetic (PM) state above about 200 K is close to the free ion value 2.54 μ_B of Ce^{3+}, indicating localized $4f$ electrons. The estimated entropy via specific heat measurements suggests that the Kondo energy scale is comparable to the RKKY interaction scale ($T_K \sim T_N$), where the f electrons strongly hybridize with conduction electrons. Recent magnetization and specific heat experiments on $CeRhAl_4Si_2$ have clarified the complex phase diagram of external field (H_0) vs temperature, and proposed that the $J = 5/2$ manifold is split into three doublets by the crystalline electric field (CEF), i.e., the ground state would be a $\Gamma_7^{(1)}$ doublet, the first excited Γ_6 doublet would lie at E_1 = 136 K, and the second

[3] Present Address: Materials Science Division, Argonne National Laboratory, Argonne, Illinois 60439, USA

excited $\Gamma_7^{(2)}$ doublet is estimated to be 342 K from the ground state. [2] Indeed, this CEF level scheme has been used to explain the *Ising*-anisotropy of magnetic susceptibility along the c-axis. Two AFM transitions at T_{N1} and T_{N2} are observed in specific heat, confirming long-range antiferromagnetism [1, 2]. Powder neutron diffraction experiments [3] provide evidence for a simple AFM structure below T_{N2} with a propagation vector $\boldsymbol{Q}_2 = (0, 0, 1/2)$, where the Ce moments of $\sim$1 μ_B align ferromagnetically in the c plane and stack antiferromagnetically along the c-axis. Recently, the intermediate AFM structure has been determined by neutron diffraction on single crystals [4] to have an incommensurate wavevector of $\boldsymbol{Q}_1 = (\delta, \delta, 1/2)$, with a temperature dependent propagation and "locks in" to the commensurate AFM at T_{N1}. To account for their data, the authors propose a model of spin-density wave order with the Ce moments modulated along the a-axis due to Fermi surface nesting. [4]

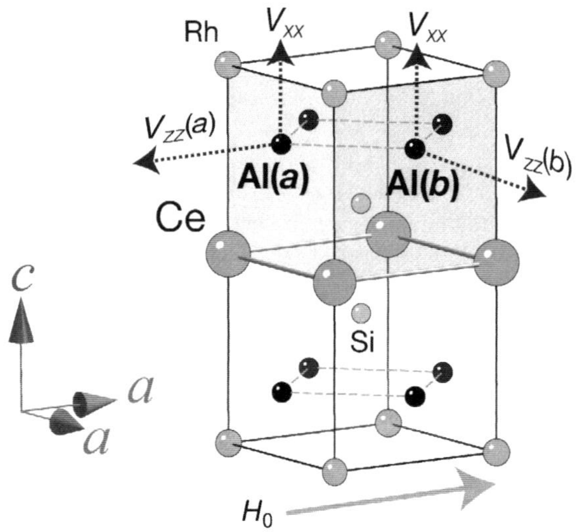

Figure 1. (Color Online) Unit cell of $CeRhAl_4Si_2$. The local principal axes of electrical field gradient on the Al sites are also shown. Although all the Al sites are crystallographically equivalent, if the external field (H_0) is applied along the a axis, two Al sites which are denoted as Al(a) and Al(b) sites are distinguished by local field directions, i.e., $H_0 \parallel V_{ZZ}$ and $H_0 \perp V_{ZZ}$, respectively.

In general, nuclear magnetic resonance (NMR) is a particularly useful tool for the microscopic investigation of the magnetism of materials, providing information about magnetic ordering that is complementary to other microscopic probes. Indeed, our spectral study of ^{27}Al NMR has confirmed the incommensurability between T_{N1} and T_{N2} and the commensurability below T_{N1}, which will be published elsewhere. In this proceedings, we will discuss anisotropy of spin fluctuations in the PM state of $CeRhAl_4Si_2$, studied through nuclear relaxation measurements.

2. Experimental

Single crystals of $CeRhAl_4Si_2$ were grown from Al/Si flux.[1] The chemical composition and homogeneity of single crystals were confirmed using a scanning electron microscope (SEM) with an energy dispersive x-ray spectrometer (EDS). A single crystal with dimensions $2 \times 2 \times 0.16$ mm^3 was inserted into an rf excitation coil of copper, and mounted on a two-axes goniometer installed in the NMR probe. NMR measurements were carried out using a phase-coherent, pulsed spectrometer. External magnetic fields were applied using a homogeneous superconducting magnet specified for NMR. The nuclear spin-lattice relaxation time T_1 was measured using the inversion-recovery method with a π-pulse. Values of T_1 were obtained from fits to an appropriate relaxation function. The magnetization recovery ($\{M(\infty) - M(t)\}/M(\infty)$) for the central and second satellite NMR transitions of the ^{27}Al ($I = 5/2$) nuclei gave satisfactory fits to the single-T_1 functions: $\frac{1}{35}\exp(-t/T_1) + \frac{8}{45}\exp(-6t/T_1) + \frac{50}{63}\exp(-15t/T_1)$ for the central transition, and $\frac{1}{35}\exp(-t/T_1) + \frac{3}{14}\exp(-3t/T_1) + \frac{2}{5}\exp(-6t/T_1) + \frac{2}{7}\exp(-10t/T_1) + \frac{1}{14}\exp(-15t/T_1)$ for the second satellite transition, respectively.

3. Results and discussions

From the ^{27}Al NMR spectral study, it has been identified that the principal electrical field gradient (EFG) axis (V_{ZZ}) is perpendicular to the ac-plane and that the V_{XX} is parallel to the c-axis, as shown in Fig. 1, reflecting the local orthorhombic symmetry ($2mm$) of the Al sites. The EFG parameters are obtained at 50 K as $\nu_{\rm Q} = \frac{3e^2qQ}{2I(2I-1)h} = 1.607$ MHz, with an asymmetry EFG parameter $\eta = \frac{|V_{YY}-V_{XX}|}{V_{ZZ}} = 0.856$. If the external field (H_0) is applied along the a-axis, as illustrated in Fig. 1, the Al sites split into two magnetically distinguished sites, i.e., H_0 is parallel to V_{ZZ} on half of the Al sites, which are defined as Al(a) sites, but H_0 is perpendicular to V_{ZZ} on the other half of Al sites, defined as Al(b) sites. Here, the b axis is defined as a local axis (in-plane) along V_{YY}.

In this proceedings, our concern is to consider the anisotropy of spin fluctuations in the PM state. In general, $1/T_1$ on the ligand sites can be written [5] as

$$\frac{1}{T_1} = 2(\gamma_{\rm N}A_{\perp}/\gamma_{\rm e})^2 T \sum_{\boldsymbol{q}} f_{\perp}^2(\boldsymbol{q}) \frac{{\rm Im}\chi_{\perp}(\boldsymbol{q},\omega_0)}{\omega_0}, \quad (1)$$

in units of $k_B = \hbar = 1$, and where $\gamma_{\rm e}$ is the electronic gyromagnetic ratio, A_α is the transferred hyperfine coupling constant, $f_\alpha(\boldsymbol{q})$ is the hyperfine form factor (taken as unity for simplicity in the following analysis), ${\rm Im}\chi(\boldsymbol{q},\omega_0)$ is the imaginary part of the dynamical susceptibility by f electrons, ω_0 is the nuclear Larmor frequency and $\perp$ refers to the component perpendicular to the quantization axis. Thus, the nuclear spin-lattice relaxation rate $1/T_1$ is driven by low-energy spin-fluctuation spectra, and $1/T_1$ is sensitive to the perpendicular fluctuations in an applied external field. In order to decompose in-plane and out-of-plane components of spin fluctuations, a new spin-lattice relaxation rate can be defined as $R_i \equiv (\gamma_{\rm N}A_i/\gamma_{\rm e})^2 \sum_{\boldsymbol{q}} \frac{{\rm Im}\chi_i(\boldsymbol{q},\omega_0)}{\omega_0}$ $(i = a, b, c)$. Using these new R_i, the experimental $1/T_1T$ along a, b, and c can be written as

$$\left(\frac{1}{T_1T}\right)_{H_0\parallel a} = R_b + R_c; \quad \left(\frac{1}{T_1T}\right)_{H_0\parallel b} = R_c + R_a; \quad \left(\frac{1}{T_1T}\right)_{H_0\parallel c} = R_a + R_b \quad (2)$$

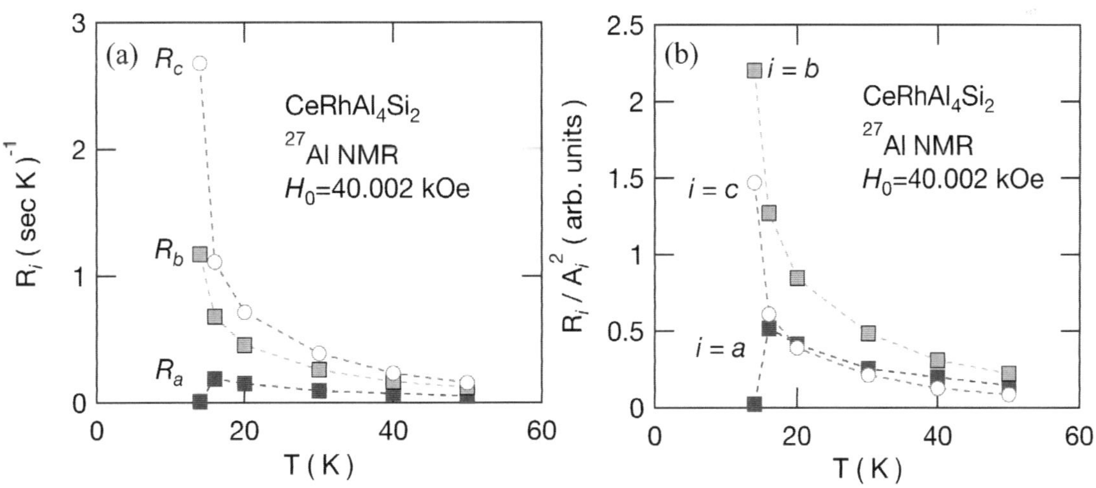

Figure 2. (Color online) (a) Directionally decomposed spin-lattice relaxation rates divided by temperature along the a-, b-, and c-axes. (b) Normalized $R_i/|A_i|^2$ ($i = a, b, c$) for ^{27}Al NMR in CeRhAl$_4$Si$_2$ under an external field of H_0=40.002 kOe.

For temperature above ~100 K, the values and temperature dependence of $1/T_1$ can be well reproduced by the computed local spin fluctuations, assuming that the f electrons are fully localized. Namely, using the proposed CEF scheme,[2] this contribution may be calculated from the following formula [6, 7, 8, 9],

$$\left(\frac{1}{T_1}\right)_{\rm CEF} = \frac{2\gamma_{\rm N}^2|(A/z')^2|}{\omega_{\rm ex}} \sum_j \frac{|\langle J_z\rangle_j|^2 \exp(-E_j)/k_{\rm B}T}{Z}, \quad (3)$$

with a characteristic exchange frequency $\omega_{\rm ex}$, where z' and z are the number of nearest neighboring Ce ions from the ligand sites and Ce sites, respectively, and $\langle J_z\rangle_j$ is the expectation value of J_z for the j-th CEF eigenstate and Z is the single-ion partition function. $\omega_{\rm ex}$ may be expressed as $(\sqrt{2z}/\sqrt{3}\hbar)J_{\rm ex}p_{\rm eff}$ with an exchange interaction $J_{\rm ex}$ and an effective moment $p_{\rm eff}$ of Ce ions. [10] The $1/T_1$ data can be reproduced by the sum of $(1/T_1)_{\rm CEF} + rT$ with $p_{\rm eff} = 2.54\ \mu_{\rm B}$, $J_{\rm ex} \sim 14$ K and a small Korringa rate r of 0.07 (sec K)$^{-1}$ due to conduction electrons (not shown). To determine the spin-fluctuation contribution of f electrons to R_i, a small contribution r of Korringa term is subtracted before the decomposition of $1/T_1T$. Below ~50 K, the deviation of $1/T_1$ from the calculated curve is seen (not shown), which may suggest development of hybridization between f and conduction electrons.

Figure 2(a) shows the temperature dependence of the derived R_i ($i = a, b, c$) from Eq. 2 below ~50 K in the PM state of $CeRhAl_4Si_2$. Although the R_i are rather isotropic around 50 K, R_c and R_b increase toward $T_{\rm N1}$ and R_a is nearly temperature independent. In order to compare the R_i quantitatively, as shown in Fig. 2(b), each R_i are divided by the square of the transferred hyperfine coupling constant A_i, which is determined as $A_a = -0.60$ kOe/$\mu_{\rm B}$, $A_b = -0.73$ kOe/$\mu_{\rm B}$, and $A_c = 1.35$ kOe/$\mu_{\rm B}$, respectively, from the Knight shift measurements. Interestingly, spin fluctuations along the b-axis become the largest and those along the c-axis also increase after the b-component with decreasing temperature. On the other hand, the a-component is less enhanced and becomes much smaller near $T_{\rm N1}$.

In general, anisotropy of $R_i/|A_i^2|$ in PM state near the ordering temperature closely reflects the direction of the ordered moments, because $1/T_1$ on the ligand sites senses the local fluctuations from the neighboring magnetic sites. In such cases, the spectrum of dynamical susceptibility can be approximated by Lorentzian (or Gaussian in some cases) form with spectral width Γ_i at a specific $\boldsymbol{q}$, which corresponds to an ordering propagation vector. Indeed, the anisotropy of $1/T_1T$ can tell the preferred orientation of ordered moments in many cases: for example, in the heavy fermion ferromagnet $CeRu_2Ga_2B$ [11], and heavy fermion antiferromagnets $CePd_5Al_2$ [12], $UPtGa_5$ [13] $NpFeGa_5$ [13], and $NpCoGa_5$ [14]. In $CeRhAl_4Si_2$, from Fig. 2(b), one might expect moments in the ordered state to lie along the b-axis. Below $T_{\rm N1}$, however, the ordered moment is identified to be along the c-axis, while an incommensurate spin density modulation $\boldsymbol{Q}^* = (\delta, \delta, 1/2)$ of the ordered moments is also identified between $T_{\rm N1}$ and $T_{\rm N2}$ by NMR and neutron diffraction measurements [4]. It is noted that such a moment modulation vanishes below $T_{\rm N2}$, and that the AFM propagation is locked to $\boldsymbol{Q} = (0, 0, 1/2)$ below $T_{\rm N2}$. [3, 4] Thus, in the case of $CeRhAl_4Si_2$, the anisotropy of R_i/A_i^2 cannot predict the orientation of ordered moments correctly.

In principle, in-plane anisotropy of R_i/A_i^2 between a- and b- directions under zero-field is not expected because the electronic ${\rm Im}\chi(\boldsymbol{q}, \omega)$ should have tetragonal symmetry. Therefore, there should be a directional dependence of the in-plane fluctuation spectrum induced by external fields, which may be due to a strong nesting instability, at $\boldsymbol{Q}_{\rm nest} = (\delta, \delta, q_z)$ with arbitrary q_z, coming from the quasi-two dimensional electronic state [1]. On the other hand, the anisotropy between c- and $a-$ directions reflect anisotropy of AFM fluctuations at $\boldsymbol{Q}$. The first order transition at $T_{\rm N2}$ from incommensurate to commensurate AFM orderings [1, 2, 4] may come from such a band effect (nesting instability) which was previously discussed theoretically [15, 16].

4. Summary

We have performed T_1 measurements for ^{27}Al NMR in the Kondo lattice compound $CeRhAl_4Si_2$. At high temperatures, the rather isotropic $1/T_1$ can be interpreted as fully localized f electrons. Below ~50 K, the f electrons hybridize with conduction electrons to acquire itinerant character. Then, $1/T_1$ begins to show an anisotropy reflecting anisotropic spin fluctuations and a nesting instability in this compound. Further Fermi surfaces investigations using quantum oscillations or angle-resolved photoemisison spectroscopy in $CeRhAl_4Si_2$ should be particularly promising.

5. Acknowledgments

We thank J. Lawrence for valuable discussions. Work at Los Alamos National Laboratory was performed under the auspices of the US Department of Energy, Office of Basic Energy Sciences, Division of Materials Sciences and Engineering. Work in Japan was partly supported by the Reimei Research Program of JAEA. section.

References

[1] Ghimire N J, Ronning F, Williams D J, Scott B L, Luo Y, Thompson J D and Bauer E D 2015 *J. Phys. :Condens. Matter* **27** 025601
[2] Maurya A, Kulkarni R, Thamizhavel A, Paudyal D and Dhar S K 2015 *cond-mat* arXiv:1501.00250
[3] Ghimire N J, Calder S, Janoschek M and Bauer E D 2015 *J. Phys. :Condens. Matter* **27** 245603
[4] Gunasekera J, Harriger L, Dahal A, Maurya A, Heitmann T, Disseler S, Thamizhavel A, Dhar S and Singh D K 2015 *cond-mat* arXiv:1509.02092
[5] Moriya T 1963 *J. Phys. Soc. Jpn.* **18** 516
[6] Tanida H, Takagi S, Suzuki H S, Satoh I and Komatsubara T 2006 *J. Phys. Soc. Jpn.* **75** 074721
[7] Moriya T and Obata Y 1958 *J. Phys. Soc. Jpn.* **13** 1333
[8] Sugawara K 1977 *J. Phys. Soc. Jpn.* **42** 1154
[9] Sugawara K 1977 *J. Phys. Soc. Jpn.* **42** 1161
[10] Moriya T 1956 *Prog. Theor. Phys.* **16** 23,641
[11] Sakai H, Tokunaga Y, Kambe S, Baumbach R E, Ronning F, Bauer E D and Thompson J D 2012 *Phys. Rev. B* **86** 094402
[12] Sakai H, Chudo H, Tokunaga Y, Kambe S, Haga Y, Ronning F, Bauer E D, Thompson J D, Homma Y, Aoki D, Nakano Y, Honda F, Settai R and Ōnuki Y 2012 *J. Phys. Soc. Jpn. Suppl. B* **81** SB003
[13] Kambe S, Sakai H, Tokunaga Y, Fujimoto T, Walstedt R E, Ikeda S, Aoki D, Homma Y, Haga Y, Shiokawa Y and Ōnuki Y 2007 *Phys. Rev. B* **75** 140509(R) (4)
[14] Sakai H, Kambe S, Tokunaga Y, Fujimoto T, Walstedt R E, Yasuoka H, Aoki D, Homma Y, Yamamoto E, Nakamura A, Shiokawa Y and Ōnuki Y 2007 *Phys. Rev. B* **76** 024410 (14)
[15] Rice T M 1970 *Phys. Rev. B* **2** 3619
[16] Hasegawa H 1978 *Journal of Low Temperature Physics* **31** 475

Photoinduced itinerant ferromagnetism in copper octacyanomolybdates

Jun Ohara and Shoji Yamamoto

Department of Physics, Hokkaido University, Sapporo 060-0810, Japan

E-mail: `ohara@phys.sci.hokudai.ac.jp`

Abstract. We make a microscopic theory of the photoswitchable magnetism in copper octacyanomolybdates. By numerically solving a time-dependent Schrödinger equation based on the relevant extended Hubbard model, we reproduce magnetization by green-light irradiation and subsequent demagnetization by orange-light irradiation. At the onset of the ferromagnetism, the charge-transfer gap disappears. In an attempt to stimulate experimental investigations, we simulate time evolution of the angle-resolved photoemission spectroscopy and optical-conductivity spectra.

1. Introduction

Photomagnetism of cyano-bridged transition metal complexes has been the central issue in both physical and chemical fields. In Fe-Co Prussian blue analogues consisting of hexacyanoferrates and cobalt ions, photoirradiation induces ferrimagnetism [1, 2, 3]. The diamagnetic entity $Fe^{II}(S=0)$–CN–$Co^{III}(S=0)$ is photoconverted to the magnetic one $Fe^{III}(S=1/2)$–CN–$Co^{II}(S=3/2)$. The electron transfer mediated by a cyano group triggers the magnetic interaction. The reverse process is also induced by the photoirradiation with longer wavelength [4]. Such photo-manipulation of magnetism by selective charge-transfer excitations is very interesting from the viewpoints of fundamental physics and its application.

Several experimental groups recently discovered photomagnetism in copper-molybdenum bimetallic assemblies $Cu_2[Mo(CN)_8]\cdot n H_2O$ [5, 6, 7]. These compounds consist of octacyanomolybdates $[Mo^{IV}(CN)_8]^{4-}$ of the square antiprism configuration [8, 9] and Cu ions. As is shown in Fig. 1(a), the crystal structure is characterized by the $I4/m$ tetragonal space group with $\mathbf{C}_{4h}$ point group. The coordination network based on octacyanomolybdates brings about various electron-transfer and magnetic-exchange pathways, in comparison with those in the Prussian blue analogues with simple cubic structures [10]. Fully-filled $4d_{z^2}$ orbitals on Mo^{IV} sites and half-filled $3d_{yz/zx}$ orbitals on Cu^{II} sites are physically active due to a crystalline-field splitting. Irradiation of the nonmagnetic ground state with a laser light of 473 nm induces a macroscopic magnetization, while subsequent irradiation with a laser light of 658 nm reduces this ferromagnetism. Further photoirradiation with longer wavelength reduces the induced magnetization step by step. This phototunable magnetism is stable from 3 K to 100 K. In the magnetization process, photoirradiation induces electron transfers from the fully-filled Mo orbitals to the half-filled Cu orbitals. As a result, both Mo and Cu ions have magnetic moments and then the ferromagnetic exchange pathway appears. These electron configurations are supposed by the measurements of the magnetization curves before and after photoirradiation [5].

However it remains to be fully elucidated whether the demagnetization can be described as a simple reverse process of the magnetization or not.

In spite of the intriguing observations, neither photoinduced magnetism nor ground-state magnetism has been interpreted yet beyond the phenomenological understanding. In order to reveal a microscopic picture of the phototunable magnetism of $Cu_2[Mo(CN)_8]$, we draw a ground-state phase diagram and simulate successive photoirradiation of the nonmagnetic ground state.

2. Modelling

In order to describe the dicopper octacyanomolybdates, we employ the 2/3-filling extended Hubbard model represented by

$$\begin{aligned}
\mathcal{H} = & \sum_{<n,m,i,j>} \sum_{\sigma,\tau} \Bigg\{ \frac{t_{\mathrm{MoCu}}}{2} \Big[(-1)^{i+j+1} c^{\dagger}_{n:\mathrm{Cu}(i)\sigma} c_{m:\mathrm{Mo}(j)\sigma} + \mathrm{H.c.} \Big] + V_{\mathrm{MoCu}} n_{n:\mathrm{Cu}(i)\sigma} n_{m:\mathrm{Mo}(j)\tau} \\
& + J_{\mathrm{MoCu}} c^{\dagger}_{n:\mathrm{Cu}(i)\sigma} c^{\dagger}_{m:\mathrm{Mo}(j)\tau} c_{n:\mathrm{Cu}(i)\tau} c_{m:\mathrm{Mo}(j)\sigma} \\
& \quad + \frac{J'_{\mathrm{MoCu}}}{2} (c^{\dagger}_{n:\mathrm{Cu}(i)\sigma} c^{\dagger}_{n:\mathrm{Cu}(i)\tau} c_{m:\mathrm{Mo}(j)\tau} c_{m:\mathrm{Mo}(j)\sigma} + \mathrm{H.c.}) \Bigg\} \\
& + \sum_{<n\neq m>} \sum_{i=1}^{4} \sum_{\sigma,\tau} \Bigg[\frac{t_{\mathrm{CuCu}}}{2} (c^{\dagger}_{n:\mathrm{Cu}(i)\sigma} c_{m:\mathrm{Cu}(i+4)\sigma} + \mathrm{H.c.}) + V_{\mathrm{CuCu}} n_{n:\mathrm{Cu}(i)\sigma} n_{m:\mathrm{Cu}(i+4)\tau} \\
& + J_{\mathrm{CuCu}} c^{\dagger}_{n:\mathrm{Cu}(i)\sigma} c^{\dagger}_{m:\mathrm{Cu}(i+4)\tau} c_{n:\mathrm{Cu}(i)\tau} c_{m:\mathrm{Cu}(i+4)\sigma} \\
& \quad + \frac{J'_{\mathrm{CuCu}}}{2} (c^{\dagger}_{n:\mathrm{Cu}(i)\sigma} c^{\dagger}_{n:\mathrm{Cu}(i)\tau} c_{m:\mathrm{Cu}(i+4)\tau} c_{m:\mathrm{Cu}(i+4)\sigma} + \mathrm{H.c.}) \Bigg] \\
& + \sum_{n,\sigma} \Bigg[\sum_{i=1}^{8} (\varepsilon_{\mathrm{Cu}} + \frac{U_{\mathrm{Cu}}}{2} n_{n:\mathrm{Cu}(i)-\sigma}) n_{n:\mathrm{Cu}(i)\sigma} + \sum_{j=1}^{4} (\varepsilon_{\mathrm{Mo}} + \frac{U_{\mathrm{Mo}}}{2} n_{n:\mathrm{Mo}(j)-\sigma}) n_{n:\mathrm{Mo}(j)\sigma} \Bigg] \qquad (1)
\end{aligned}$$

with $n_{n:A(i)} = \sum_{\sigma} n_{n:A(i)\sigma} = \sum_{\sigma} c^{\dagger}_{n:A(i)\sigma} c_{n:A(i)\sigma}$, where $c^{\dagger}_{n:A(i)\sigma}$ creates an electron of spin $\sigma = \uparrow, \downarrow \equiv \pm$ on the Cu $3d$ (A = Cu; i = 1-8) or Mo $4d$ (A = Mo; i = 1-4) orbitals labelled i at unit cell n, as is shown in Fig. 1(a). $\sum_{<n,m,i,j>}$ and $\sum_{<n\neq m>} \sum_{i=1}^{4}$ run all over the nearest neighboring Cu-Mo bonds and Cu-Cu bonds, respectively. $t_{\mathrm{MoCu/CuCu}}$, $V_{\mathrm{MoCu/CuCu}}$, $J_{\mathrm{MoCu/CuCu}}$, and $J'_{\mathrm{MoCu/CuCu}}$ describe transfer integrals, intermetallic Coulomb repulsions, exchange interactions and pair-hopping interactions, respectively. On-site Coulomb repulsions $U_{\mathrm{Mo/Cu}}$ and orbital energies of isolated metals $\varepsilon_{\mathrm{Mo/Cu}}$ are considered. All the two-body terms are treated within a Hartree-Fock (HF) scheme.

3. Ground-state properties

Based on the symmetry argument on the crystal structure [11, 12], we obtain possible electronic states as is shown in Fig. 1(b). Both nonmagnetic and magnetic states appear as a ground state. The charge-disproportionate (CDP) and ferromagnetic (FM) states preserve the original translational symmetry property, while the charge-density-wave (CDW) and antiferromagnetic (AFM) states show cell-doubling structures. There are two kinds of charge-disproportionate patterns. Mo-CDP can take the $\mathrm{Mo^{IV}}$-$\mathrm{Cu^{II}}$ valence arrangement suggested by the observations [5, 7], while Cu-CDP exhibits electron localization in Cu sites. Only FM has a global magnetization.

Figure 1(c) shows a ground-state phase diagram at the sufficiently low temperature $k_{\mathrm{B}}T = 0.001t_{\mathrm{MoCu}}$, where we set $t_{\mathrm{CuCu}} = 0.3t_{\mathrm{MoCu}}$, $V_{\mathrm{CuCu}} = 2.4t_{\mathrm{MoCu}}$, $V_{\mathrm{MoCu}} = 0.8t_{\mathrm{MoCu}}$, $\varepsilon_{\mathrm{Cu}} - \varepsilon_{\mathrm{Mo}} = 1.0t_{\mathrm{MoCu}}$, and $J_{\mathrm{MoCu}} = J'_{\mathrm{MoCu}} = -J_{\mathrm{CuCu}} = -J'_{\mathrm{CuCu}} = 0.6t_{\mathrm{MoCu}}$ unless otherwise noted.

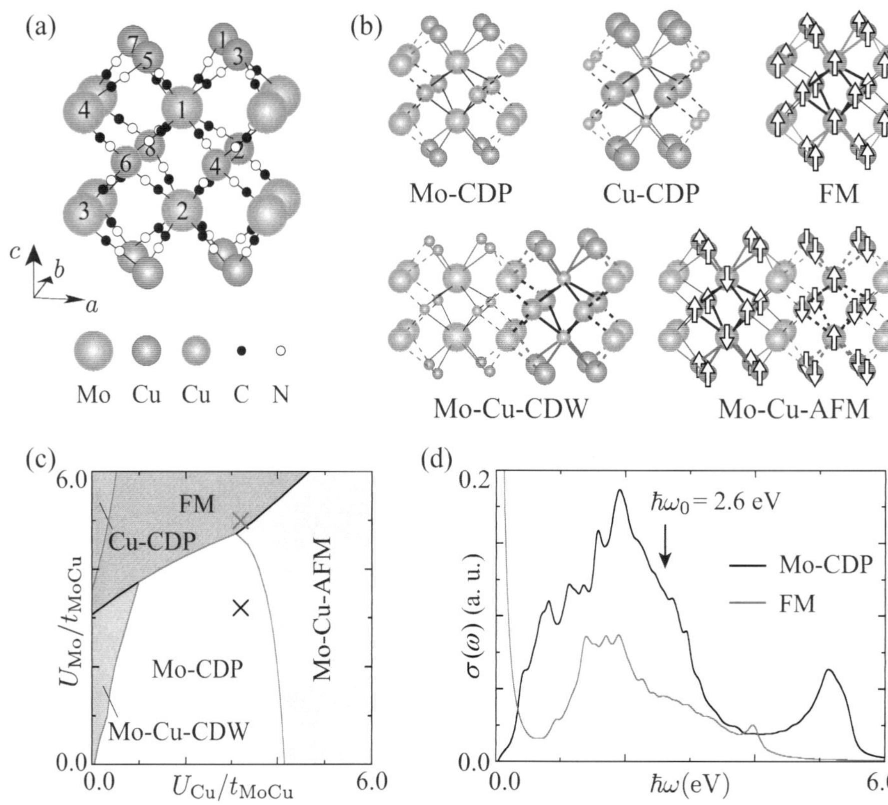

Figure 1. (Color online) (a) A crystal structure of $Cu_2[Mo(CN)_8]$. A unit cell contains four Mo ions and eight Cu ions. (b) Possible electronic states, where varying circles indicate electron densities and white arrows describe electron spins. (c) A ground-state phase diagram as a function of on-site Coulomb repulsions. Black and red cross symbols (×) point to (U_{Cu}/t_{MoCu}, U_{Mo}/t_{MoCu})=(3.2, 3.2) and (3.2, 5.0), respectively. (d) Optical conductivity spectra of the Mo-CDP and FM states, where a vertical arrow indicates a pumping photon energy.

Phase boundaries drawn in black and red denote transitions of the first order and the second order, respectively. With increasing U_{Cu} and U_{Mo}, FM and Mo-Cu-AFM are stabilized. These magnetic phases are strongly affected by the direct exchange interactions J_{MoCu} and J_{CuCu} as well. In the present calculation, we chose positive J_{MoCu} stabilizing ferromagnetism because the analogue compound $Cu_{1.5}[Mo(CN)_8]\cdot 3H_2O$ shows a ferromagnetic correlation between spins on Mo and Cu sites [7]. On the other hand, J_{CuCu} is taken to be negative. Here we note the identity $\sum_{\sigma,\tau} c^{\dagger}_{n:A(i)\sigma} c^{\dagger}_{m:A'(i')\tau} c_{n:A(i)\tau} c_{m:A'(i')\sigma} = -2 s_{n:A(i)} \cdot s_{m:A'(i')} - n_{n:A(i)} n_{m:A'(i')}/2$, where $s^{\lambda}_{n:A(i)} = \sum_{\sigma,\sigma'} c^{\dagger}_{n:A(i)\sigma} c_{n:A(i)\sigma'} \tau^{\lambda}_{\sigma\sigma'}/2\,(\lambda = x, y, z)$ and τ^{λ} indicates Pauli spin matrices. In a weak correlation regime, Mo-Cu-CDW is most stable due to the relevant intermetallic Coulomb interactions. Mo-CDP predominates in a moderate correlation regime. Assuming that both U_{Cu} and U_{Mo} values of $Cu_2[Mo(CN)_8]$ are in this regime, we chose $U_{Cu} = U_{Mo} = 3.2t_{MoCu}$ as standard parametrization, where the valence arrangement of Mo and Cu ions agrees with the observations [11, 12]. Cu-CDP will never appear as a ground state of $Cu_2[Mo(CN)_8]$ because it is stabilized only in an unrealistic parameter region as $U_{Mo} >> U_{Cu}$. Mo-CDP and FM are sitting adjacently on the phase diagram, and therefore, fluctuation by external fields is likely to induce the phase transition between them.

We further calculate optical conductivities of Mo-CDP and FM. In the FM state, we set $(U_{\rm Cu}/t_{\rm MoCu}, U_{\rm Mo}/t_{\rm MoCu}) = (3.2, 5.0)$. Hereafter the intermetallic transfer integral $t_{\rm MoCu}$ is set equal to 0.7 eV, which is comparable to that in typical Prussian blue photomagnets [13, 14]. The optical conductivity spectrum is defined as

$$\sigma(\omega) = -\frac{\pi}{N}\left(\langle E_0|\mathcal{T}|E_0\rangle + 2\sum_{l\neq 0}\frac{|\langle E_l|\mathcal{J}|E_0\rangle|^2}{E_l - E_0}\right)\delta(\omega) + \frac{\pi}{N\omega}\sum_{l\neq 0}|\langle E_l|\mathcal{J}|E_0\rangle|^2\delta(E_l - E_0 - \hbar\omega), \quad (2)$$

where $|E_l\rangle$ is a one-electron-hole-pair excited state from $|E_0\rangle$ being either Mo-CDP or FM. $\mathcal{J}$ and $\mathcal{T}$ indicate a current operator and a kinetic energy operator, respectively [15, 16]. Considering the electric current in the a-b plane [12], we obtain the optical conductivity perpendicular to the c axis as is shown in Fig. 1(d). Mo-CDP shows charge-transfer gap $\Delta = 0.35$ eV. A broad absorption band is attributed to the electron excitations from the highest-occupied molecular orbital (HOMO) to the lowest-unoccupied molecular orbital (LUMO) over the whole Brillouin zone. On the other hand, FM shows a significant zero-frequency absorption namely Drude component. The FM state exhibits a metallic conductivity. Thus we find that the insulating Mo-CDP closely competes with the itinerant FM at low temperatures. An electric conductivity serves as a probe of the magnetic transition in $Cu_2[Mo(CN)_8]$.

4. Photomagnetization

Since photomagnetization is hardly realized at a zero magnetic filed, we apply a moderate filed to the system as follow,

$$\mathcal{H}_{\rm ex} = -g\mu_{\rm B}H\sum_n\left[\sum_{i=1}^{8} s^z_{n:{\rm Cu}(i)} + \sum_{j=1}^{4} s^z_{n:{\rm Mo}(j)}\right], \quad (3)$$

where $H = 2.5$ T. Then down-spin electrons are selectively photoexcited. Such a photoexcitation is described by multiplying the hopping term $c^\dagger_{n:A(i)-}c_{n':A'(i')-}$ by the Peierls phase factor $\exp[-({\rm i}e/\hbar v)A(t)\cdot(r_{n:A(i)} - r_{n':A'(i')})]$ [17], where e, v, and $r_{n:A(i)}$ are the elementary charge, the light velocity, and the position vector, respectively. We consider successive photoirradiation described by the vector potential

$$A(t) = A_0\sum_{i=0}^{3} e^{-\gamma^2(t-t_i)^2}\cos\omega_i t, \quad (4)$$

where centers and photon energies of first~fourth ($i = 0 \sim 3$) pulses are given by $t_i = 0.38 + 0.75i$ ps and $\hbar\omega_i = (1 - 0.25i)\hbar\omega_0$, respectively. With the experimental observations in mind [7], we set $\hbar\omega_0$ equal to 2.6 eV, which is indicated by a vertical arrow in Fig. 1(d). A width and an amplitude of each pulse are $2\gamma^{-1} = 0.25$ ps and $|A_0| = A_0 = 2.0(\hbar v/ea)$, respectively. Furthermore, we consider Dzyaloshinskii-Moriya (DM) interaction breaking the conservation law for the total magnetization:

$$\mathcal{H}_{\rm DM} = \sum_n\sum_{l=1}^{4}\sum_{\rho=0}^{1}\sum_{\sigma,\sigma'=0}^{1}(-1)^{\rho+\sigma}D^{(\sigma'\sigma)}_{l+4\rho}[s_{n+\delta(l,\rho,\sigma,\sigma'):{\rm Mo}(1+3\sigma'+(-1)^{\sigma'}\rho)} \times s_{n:{\rm Cu}(2l-\sigma)}], \quad (5)$$

where $\delta(l,\rho,\sigma,\sigma') = (\sigma'{\rm Re}[f(l)], \sigma'{\rm Im}[f(l)], \sigma\rho)$ with $f(l) = e^{{\rm i}\pi/4}[1 + e^{{\rm i}\pi(1-l)/2}]/\sqrt{2}$. The DM vectors are given as $D^{(\sigma'\sigma)}_i = g_i \cdot D^{(\sigma'\sigma)}_1$, where $g_i(\in \mathbf{C}_{4h}) = C^{-1}_{4z}, C_{2z}, C_{4z}, \sigma_h, IC_{4z}, I, IC^{-1}_{4z}$ for $i = 2$ to 8, respectively, and $D^{(\sigma'\sigma)}_1 = (D, 0, 0)$ with $D = 0.7$ meV. Time evolution of the

IOP Conf. Series: Journal of Physics: Conf. Series **868** (2017) 012013 doi:10.1088/1742-6596/868/1/012013

electronic wave functions is obtained by integrating the time-dependent Schrödinegr equations step by step as

$$|\psi_{k,\nu}(t+\Delta t)\rangle = \exp[-\frac{i}{\hbar}\mathcal{H}_{\rm HF}(t)\Delta t]|\psi_{k,\nu}(t)\rangle, \quad (6)$$

where the instantaneous Hamiltonian $\mathcal{H}_{\rm HF}(t)$ consists of $\mathcal{H}$, $\mathcal{H}_{\rm ex}$, and $\mathcal{H}_{\rm DM}$ within the HF approximation [18]. The time t is discretized by $\Delta t = 0.005\hbar/t_{\rm MoCu}$. The initial state $|\psi_{k,\nu}(0)\rangle$ is the HF eigenstate of the static Hamiltonian $\mathcal{H}$, which is specified by a momentum and a band index (k,ν). The many-body wave function is given by $|\Psi(t)\rangle = \Pi_k \Pi_{\nu=1}^{16} \otimes |\psi_{k,\nu}(t)\rangle$. We irradiate Mo-CDP with a-axis-polarized light.

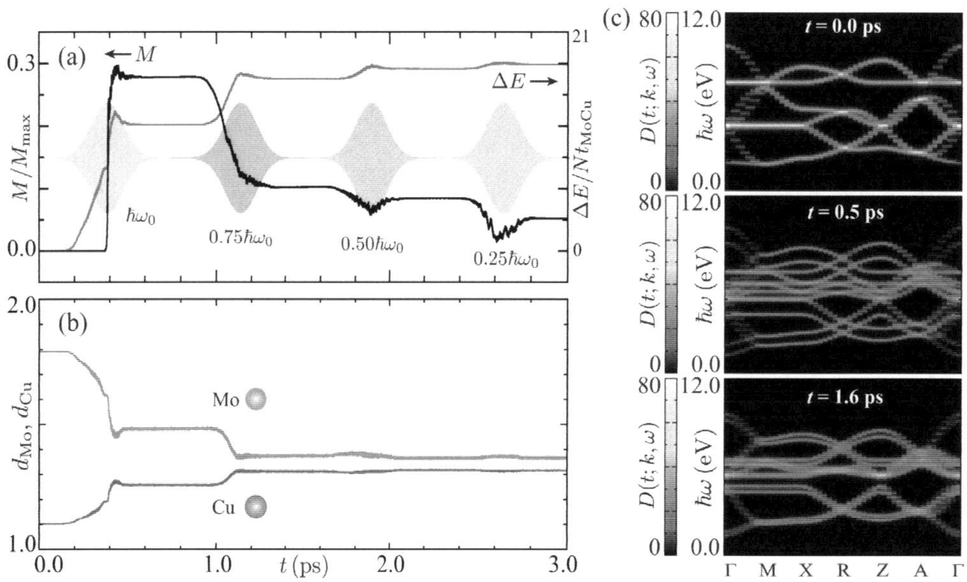

Figure 2. (Color online) (a) Time evolution of a global magnetization M and an absorbed energy ΔE. Pumping pulses are also presented. (b) Time evolution of electron densities on Mo and Cu sites. (c) Time-resolved one-particle excitation spectra at $t = 0.0, 0.5, 1.6$ ps.

We present time evolution of a global magnetization M and an absorbed photon energy ΔE in Fig. 2(a), where $M = \sqrt{\sum_\lambda (M^\lambda)^2}$ with $M^\lambda = \sum_n \langle\Psi(t)|[\sum_{i=1}^{8} s^\lambda_{n:{\rm Cu}(i)} + \sum_{i=1}^{4} s^\lambda_{n:{\rm Mo}(i)}]|\Psi(t)\rangle/N$ and $\Delta E = \langle\Psi(t)|\mathcal{H}_{\rm HF}(t)|\Psi(t)\rangle - \langle\Psi(0)|\mathcal{H}_{\rm HF}(0)|\Psi(0)\rangle$. The electronic energy increases step by step by the successive photoirradiation. Upon the first photoirradiation with $\hbar\omega_0$, ΔE reads as a double-stepped function of time. At around $t = 0.38$ ps corresponding to the pulse center, the photon absorption is suddenly accelerated. The occurrence of a significant magnetization looks simultaneous with the photon absorption at the second step of ΔE. An incubation time is needed for the rise of ferromagnetism. The second, third, and fourth photoirradiation with longer wavelength reduce the thus-induced magnetization. This step-by-step demagnetization also consists with the observations [7].

Figure 2(b) shows an electron transfer between Mo and Cu sites induced by the photoirradiation, where $d_{\rm Cu} = \sum_{i=1}^{8}\sum_n \langle\Psi(t)|n_{n:{\rm Cu}(i)}|\Psi(t)\rangle/8N$ and $d_{\rm Mo} = \sum_{i=1}^{4}\sum_n \langle\Psi(t)|n_{n:{\rm Mo}(i)}|\Psi(t)\rangle/4N$. Upon the appearance of the global magnetization, electrons transfer from Mo to Cu sites. These electrons never come back to Mo sites in the subsequent demagnetization processes. This means that the demagnetization can not be interpreted as a simple reverse process of the magnetization. The electron densities on Mo and Cu sites seem to

IOP Conf. Series: Journal of Physics: Conf. Series **868** (2017) 012013 doi:10.1088/1742-6596/868/1/012013

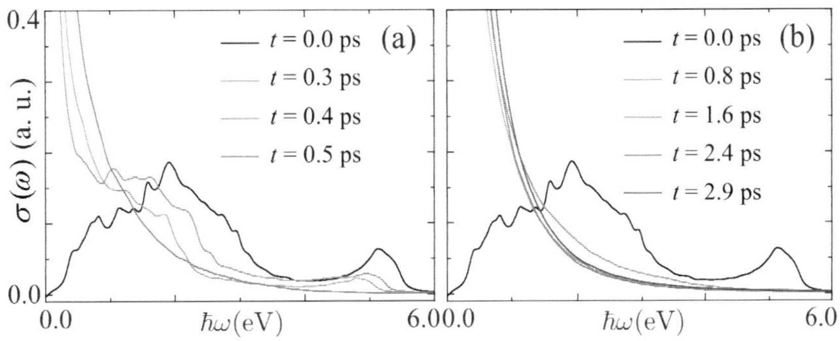

Figure 3. (Color online) Time-resolved optical conductivity spectra perpendicular to the c axis during magnetization (a) and demagnetization (b).

converge on the average value 16/12=1.33. The final steady state showing no global magnetization differs from the initial nonmagnetic state.

By calculating one-particle excitation spectra corresponding to angle-resolved photoemission spectroscopy (ARPES) spectra [19], we further investigate a microscopic electronic structure. It is given by $D(t;k,\omega) = H(t;k,\omega) + E(t;k,\omega)$ with $E(t;k,\omega) = \sum_{k,\nu}\delta(\hbar\omega - \varepsilon_{k,\nu})\langle\Psi(t)|\tilde{c}^{\dagger}_{k,\nu}\tilde{c}_{k,\nu}|\Psi(t)$ and $H(t;k,\omega) = \sum_{k,\nu}\delta(\hbar\omega - \varepsilon_{k,\nu})\langle\Psi(t)|\tilde{c}_{k,\nu}\tilde{c}^{\dagger}_{k,\nu}|\Psi(t)\rangle$, where $\tilde{c}^{\dagger}_{k,\nu}$ creates an electron in the eigenstate of energy $\varepsilon_{k,\nu}$ for the instantaneous Hamiltonian $\mathcal{H}_{\rm HF}(t)$. Figure 2(c) presents $D(t;k,\omega)$ in the photoinduced magnetization and demagnetization processes. The upper panel of Fig. 2(c) shows the band structure of the initial Mo-CDP, where the spin-up and -down electron bands are degenerate. After the first photoirradiation, this degeneracy is so lifted as to induce a net magnetization, as is shown in the middle panel of Fig. 2(c). This is the cooperative effect of the photoexcitation and the spin-mixing due to the DM interaction. After the second photoirradiation, the up- and down-spin bands converge again with a decrease in the magnetization, as is shown in the lower panel of Fig. 2(c). Upon the third and fourth photoirradiation, the band splitting further decreases with decreasing magnetization. In the demagnetization process, we observe photon emission as well as photon absorption. A resonance energy of this stimulated emission of photons also decreases with decreasing magnetization. Therefore, modulation of pumping photon energy is a key ingredient for the step-by-step demagnetization.

We further present optical conductivity spectra at several times in Fig 3. Drude component is clearly observed at t = 0.3 ps (before the appearance of the global magnetization) [see Fig. 3(a)]. With an increase in photocarriers, the initial absorption band weakens and the metallic conductivity predominates. Given that the major absorption in Mo-CDP is attributed to the HOMO-to-LUMO electron excitations, this optical feature is reasonable. The magnetic phase transition reads as an insulator-to-metal phase transition, where the metallization slightly precedes the magnetization. In the demagnetization processes, the induced metallic conductivity is almost unchanged [see Fig. 3(b)], while the induced magnetization is pronouncedly reduced [see Fig. 2(a)]. Such a nonmagnetic metallic state never appears in the ground-state phase diagram. The final steady state may thus be quite different from any equilibrium state.

5. Summary

We have demonstrated the tunable photomagnetism of $Cu_2[Mo(CN)_8]$. In the formation process of the itinerant ferromagnetism, it is within a few subpicoseconds after metallization that a global magnetization appears. This delay time may be determined by the magnitude of the DM vector giving a spin-mixing effect, where we should be reminded that $|D| \sim$meV. The

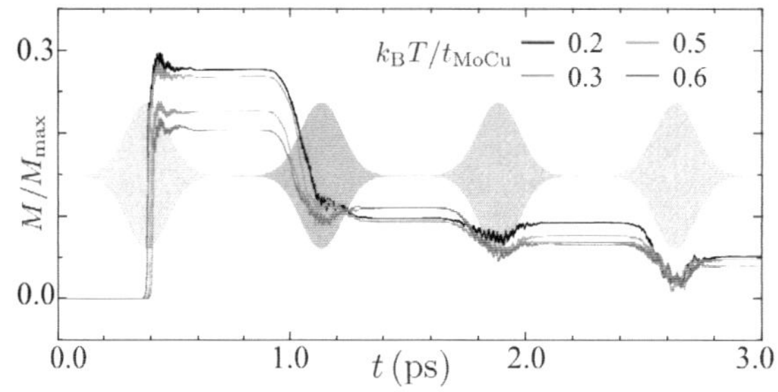

Figure 4. (Color online) Temperature dependence of the photoinduced magnetization dynamics.

demagnetization originates from stimulated emission of photons, where electrons on Mo sites further transfer to Cu sites. The photoswitching of magnetization does not necessarily mean a round trip of electrons. The initial valence arrangement is thus photoconverted to the averaged one. The final steady state is nonmagnetic but metallic, which has never been observed in thermal equilibrium.

The magnetization dynamics is almost unchanged up to $k_{\rm B}T = 0.2t_{\rm MoCu}$. As is shown in Fig. 4, above this temperature the induced magnetization is monotonically suppressed with increasing temperature. On the other hand, the step-by-step demagnetization process much less depends on temperature. The suppression of the induced magnetization is closely related to the charge-transfer gap of the ground-state Mo-CDP, $\Delta = 0.5t_{\rm MoCu}$(=0.35 eV). Thermal demagnetization effect may become remarkable above the temperature corresponding to the gap Δ. The induced magnetizations at $k_{\rm B}T = 0.3t_{\rm MoCu}$ and $k_{\rm B}T = 0.5t_{\rm MoCu}$ are indeed much different. In the demagnetization process, there are many holes in the valence band and then there is no longer any threshold temperature.

Our findings suggest that the copper octacyanomolybdates must be a functional material in that both magnetism and itinerancy of electrons are phototunable. The spin-charge dynamics will be elucidated by ultrafast pump-probe experiments, such as time-resolved ARPES [20] and conductivity [21] measurements. We hope that the present work will lead to a new insight into photoinduced dynamics in octacyanometalate-based magnets.

Acknowledgments

We are grateful to S. Ohkoshi and H. Tokoro for variable comments and fruitful discussion. This work was supported by the Ministry of Education, Culture, Sports, Science and Technology of Japan.

References

[1] O. Sato, T. Iyoda, A. Fujishima, and K. Hashimoto, Science **272**, 704 (1996).
[2] A. Bleuzen, C. Lomenech, V. Escax, F. Villain, F. Varret, C. C. dit Moulin, and M. Verdaguer, J. Am. Chem. Soc. **122**, 6648 (2000).
[3] N. Shimamoto, S. Ohkoshi, O. Sato, and K. Hashimoto, Inorg. Chem. **41**, 678 (2002).
[4] O. Sato, Y. Einaga, T. Iyoda, A. Fujishima, and K. Hashimoto, J. Electrochem. Soc. **144**, L11 (1997).
[5] S. Ohkoshi, N. Machida, Z. J. Zhong, and K. Hashimoto, Synth. Met. **122**, 523 (2001).
[6] G. Rombaut, C. Mathonière, P. Guionneau, S. Golhen, L. Ouahab, M. Verelst, and P. Lecante, Inorg. Chim. Acta **326**, 27 (2001).
[7] S. Ohkoshi, H. Tokoro, T. Hozumi, Y. Zhang, K. Hashimoto, C. Mathonière, I. Bord, G. Rombaut, M. Verelst, C. C. dit Moulin, and F. Villain, J. Am. Chem. Soc. **128**, 270 (2006).
[8] R. A. Pribush and R. D. Archer, Inorg. Chem. **13**, 2556 (1974).
[9] J. K. Burdett, R. Hoffmann, and R. C. Fay, Inorg. Chem. **17**, 2553 (1978).

[10] M. Verdaguer, A. Bleuzen, V. Marvaud, J. Vaissermann, M. Seuleiman, C. Desplanches, A. Scuiller, C. Train, R. Garde, G. Gelly, C. Lomenech, I. Rosenman, P. Veillet, C. Cartier, and F. Villain, Coord. Chem. Rev. **190-192**, 1023 (1999).
[11] J. Ohara and S. Yamamoto, J. Korean Phys. Soc. **62**, 1797 (2013).
[12] J. Ohara and S. Yamamoto, Sci. World J. **2014**, 762765 (2014).
[13] T. Kawamoto, Y. Asai, and S. Abe, Phys. Rev. B **60**, 12990 (1999).
[14] K. Yoshizawa, F. Mohri, G. Nuspl, and T. Yamabe, J. Phys. Chem. B **102**, 5432 (1998).
[15] J. Wagner, W. Hanke, and D. J. Scalapino, Phys. Rev. B **43**, 10517 (1991).
[16] D. J. Scalapino, S. R. White, and S. C. Zhang, Phys. Rev. Lett. **68**, 2830 (1992).
[17] A. Terai and Y. Ono, Prog. Theor. Phys. Suppl. **113**, 177 (1993).
[18] S. Yamamoto, J. Phys. Soc. Jpn. **80**, 084713 (2011).
[19] Y. Kanamori, J. Ohara, and S. Ishihara, Phys. Rev. B **86**, 045137 (2012).
[20] F. Schmitt, P. S. Kirchmann, U. Bovensiepen, R. G. Moore, L. Rettig, M. Krenz, J.-H. Chu, N. Ru, L. Perfetti, D. H. Lu, M. Wolf, I. R. Fisher, and Z.-X. Shen, Science **321** (2008) 1649.
[21] C. Kübler, H. Ehrke, R. Huber, R. Lopez, A. Halabica, R. F. Haglund Jr., and A. Leitenstorfer, Phys. Rev. Lett. **99**, 116401 (2007).

IOP Conf. Series: Journal of Physics: Conf. Series **868** (2017) 012014 doi:10.1088/1742-6596/868/1/012014

NMR and NQR studies of URu_2Si_2 and isostructural nonmagnetic references

Naoya Emi[1], Ryosuke Hamabata[1], Takehide Koyama[1], Koichi Ueda[1], Takeshi Mito[1], Yoh Kohori[2], Yuji Matsumoto[3], Yoshinori Haga[4], Etsuji Yamamoto[4] and Zachary Fisk[4,5]

[1]Graduate School of Material Science, University of Hyogo, Kamigori, Hyogo 678-1297, Japan
[2]Graduate School of Science, Chiba University, Chiba 263-8522, Japan
[3]Graduate School of Engineering, Nagoya Institute of Technology, Nagoya 466-8555, Japan
[4]Advanced Science Research Center, Japan Atomic Energy Agency, Tokai, Ibaraki 319-1195, Japan
[5]Department of Physics and Astronomy, University of California, Irvine, California 92697, USA

E-mail: mito@sci.u-hyogo.ac.jp

Abstract. The non-f compound $ThRu_2Si_2$ has been studied as an isostructural nonmagnetic reference for URu_2Si_2 using nuclear magnetic resonance (NMR) measurement. The temperature dependences of Knight shifts measured by ^{29}Si-NMR and ^{99}Ru-NMR are independent of temperature. The results are consistent with data previously reported on the susceptibility. ^{101}Ru-NQR frequency $^{101}\nu_Q$ was also estimated from the ^{99}Ru-NMR measurement. $^{101}\nu_Q$ of $ThRu_2Si_2$ is close to that of URu_2Si_2, especially at high temperatures, suggesting that U ions in URu_2Si_2 are in a nearly tetravalent state.

1. Introduction

The uranium heavy fermion URu_2Si_2 undergoes a mysterious phase transition, so-called "Hidden order (HO)", at $T_{HO} = 17.5$ K and an unconventional superconducting transition at $T_c = 1.4$ K [1, 2, 3]. Although intensive studies have been made on this compound, the mechanisms of these intriguing phenomena are still unclear.

To obtain detailed information on electronic states controlling physical properties in URu_2Si_2, it is very useful to compare this compound with nonmagnetic references experimentally. In our previous paper [4], we reported the results of nuclear magnetic resonance (NMR) and nuclear quadrupole resonance (NQR) measurements on two isostructural non-f compounds $ThRu_2Si_2$ and $LaRu_2Si_2$, and the comparison of the obtained results with already known data on URu_2Si_2 [5, 6] allowed us to extract new information on U valence and spin-fluctuation character of URu_2Si_2 [4]. However, it seems that the reference compounds, especially $ThRu_2Si_2$, have not been fully experimentally investigated so far. Therefore we show the first Knight shift data, to our knowledge, on $ThRu_2Si_2$ in this paper.

2. Experimental details

High quality single crystals of $ThRu_2Si_2$ were grown in a tetra-arc furnace under high-purity Ar gas atmosphere by the Czochralski method. The details of the sample preparation are described elsewhere [7]. The single crystalline sample was used for the measurement of ^{99}Ru-NMR. We also used a collection of many small single crystals for ^{101}Ru-NQR and ^{29}Si-NMR measurements. All the NMR and NQR experiments were carried out by a spin-echo technique with a phase-coherent pulsed spectrometer.

3. Results and Discussion

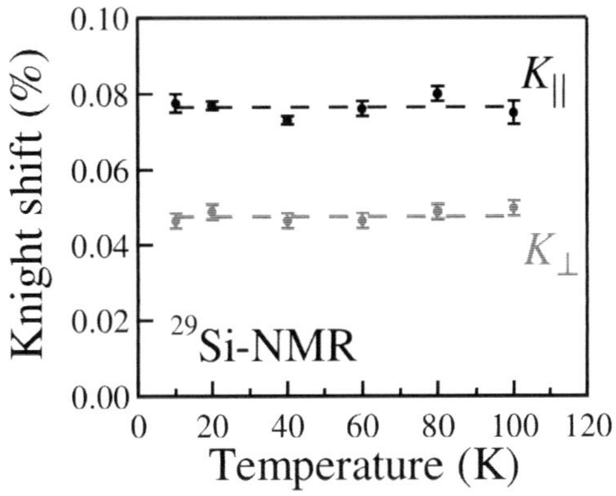

Figure 1. Temperature dependences of $^{29}K_{\parallel}$ and $^{29}K_{\perp}$ of $ThRu_2Si_2$ measured by ^{29}Si-NMR at a field of 6.0 T. The broken lines are guides to the eye.

Figure 1 shows the temperature dependences of the Knight shifts measured by ^{29}Si-NMR in $ThRu_2Si_2$. $^{29}K_{\parallel}$ and $^{29}K_{\perp}$ indicate Knight shift components for magnetic fields parallel and perpendicular to the crystal c axis, respectively. The Knight shift is generally in a linear relationship with the susceptibility χ. Therefore the facts that these Knight shifts, especially $^{29}K_{\parallel}$, show much smaller value than those for URu_2Si_2 and that they are almost independent of temperature are in good agreement with the temperature dependence of χ [8]. The anisotropic part of the shift, $^{29}K_{\text{aniso}} = \left({}^{29}K_{\parallel} - {}^{29}K_{\perp} \right)/3$, is $\sim 1 \times 10^{-2}\%$, which is almost consistent with our previous report [4]. The data in Fig. 1 are somewhat scattered, which is owing to slightly ambiguous line shapes. Since our sample was not a fine powder, the observed line shape could slightly deviate from an ideal powder pattern.

Next we show the results of NMR measurements at the Ru site. Our single crystal has a cone shape (see the upper panel of Fig. 2). The sample was mounted in a coil and aligned by eye so that the c axis is nearly perpendicular to field $\boldsymbol{B}_0$. Since nuclear spin $I = 5/2$ for ^{99}Ru, we observed a central line corresponding to the $-1/2 \leftrightarrow 1/2$ transition and nuclear quadrupole-split lines. In Fig. 2, the right panel shows the central line and the left panel shows the first satellite line corresponding to the $1/2 \leftrightarrow 3/2$ transition. To extract information on the Knight shift from the data, we used the following nuclear spin Hamiltonian;

$$\mathcal{H} = \gamma\hbar \boldsymbol{B}_0 \cdot (\widetilde{\boldsymbol{1}} + \widetilde{\boldsymbol{K}}) \cdot \boldsymbol{I} + \mathcal{H}_Q, \quad (1)$$

where γ is the gyromagnetic ratio, $\widetilde{\boldsymbol{K}}$ is the Knight shift tensor, and $\mathcal{H}_Q$ is the Hamiltonian describing nuclear quadrupole interaction. The crystal structure of $ThRu_2Si_2$ possesses a fourfold symmetry around the c axis, and $\mathcal{H}_Q$ is given by

$$\mathcal{H}_Q = (h\nu_Q/6)\left(3I_z^2 - I^2\right), \quad (2)$$

IOP Conf. Series: Journal of Physics: Conf. Series **868** (2017) 012014 doi:10.1088/1742-6596/868/1/012014

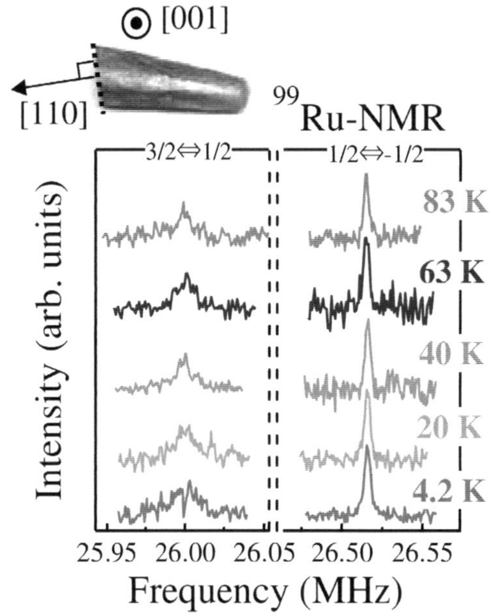

Figure 2. Upper panel: Single crystalline sample of $ThRu_2Si_2$ used for the present experiments. Lower left and right panels show the first satellite line and the central line of ^{99}Ru-NMR, respectively. The spectra were measured by applying magnetic fields of $\sim$ 13.5 T nearly perpendicular to the c axis.

where $z \parallel c$. Since NQR frequency $^{99}\nu_Q$ at 4.2 K is estimated to be 1.046 MHz from the observation of ^{101}Ru-NQR signal in zero field [4, 9], we can fit the spectra at 4.2 K to Eq. (1) with two fit parameters: an angle θ between the c axis and $\boldsymbol{B}_0$ and the Knight shift $K(\theta)$ along $\boldsymbol{B}_0$. The fit gives $\theta = 85 \pm 0.5$ degrees. Since θ should be fixed throughout the measurement, the spectra from 20 to 83 K can be fit with two fit parameters $K(\theta, T)$ and $^{99}\nu_Q(T)$. The results are shown in Fig. 3(a). The value of $^{99}K_\perp$ in Fig. 3(a) was obtained by correcting $K(\theta)$ using $^{99}K_{\rm aniso} = -0.05$ % and a relation of $K(\theta) = K_\parallel \cos^2\theta + K_\perp \sin^2\theta$. Here, $^{99}K_{\rm aniso}$ was estimated by fitting a powder pattern spectrum measured at 4.2 K (not shown), and we neglected the temperature dependence of $^{99}K_{\rm aniso}$ for the estimation of $^{99}K_\perp$ above 20 K.

The Knight shifts for both Si and Ru sites are anisotropic, $^{29}\left(K_\perp/K_\parallel\right) \sim 0.61$ and $^{99}\left(K_\perp/K_\parallel\right) \sim 1.3$, respectively, however their anisotropy are not as large as in χ ($\chi_\perp = 8.5 \times 10^{-5}$ and $\chi_\parallel < 10^{-11}$ emu/mol [8]). Generally, χ in nonmagnetic metals is a sum of diamagnetic, Van Vleck, and Pauli paramagnetic (p.p.) susceptibilities. Note that spin lattice relaxation rate $1/T_1$ at the Si site shows an isotropic Korringa relation [4], *i.e.* $^{29}(1/T_1T)_\parallel = {}^{29}(1/T_1T)_\perp$, indicating predominant isotropic coupling between Si nuclei and conduction electrons. This also suggests that considerable isotropic part of ^{29}K arises from the p.p.-susceptibility. Therefore one naively expects the existence of non-negligible diamagnetic component as well as the isotropic p.p.-susceptibility to reproduce the extremely small $\chi_\parallel$ [8].

The analysis of the data in Fig. 2 also gives the estimation of $^{101}\nu_Q$ at the Ru site. We plot in Fig. 3(b) the temperature dependence of $^{101}\nu_Q$ for $ThRu_2Si_2$ together with that of URu_2Si_2 [10]. The weak temperature dependence for $ThRu_2Si_2$ is in contrast to the data on URu_2Si_2. The anomalous temperature dependence for URu_2Si_2 is ascribed to temperature-induced changes in electronic configuration which is accompanied by a change in the U valence. Note that $^{101}\nu_Q$'s

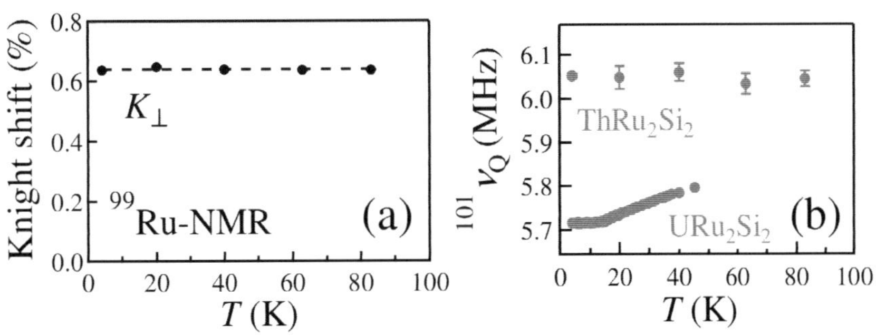

Figure 3. (a) Temperature dependence of $^{99}K_\perp$ of $ThRu_2Si_2$. (b) Temperature dependence of ν_Q of $ThRu_2Si_2$ together with the data of URu_2Si_2 [10]. Both data for $ThRu_2Si_2$ were estimated by fitting the ^{99}Ru-NMR spectra shown in Fig. 2 to Eq. (1). $^{101}\nu_Q$ was transformed from $^{99}\nu_Q$ (see Ref. [9])

of the two compounds are close to each other, especially at high temperatures. Generally ν_Q is influenced by changes in local charge distribution around the nucleus. For lanthanide based compounds, there are several examples showing that ν_Q at ligand sites significantly varies reflecting the valence change of lanthanide ions (for example see Ref. [4]). Therefore, the result may suggest that the valence of U ions at high temperature is close to Th valence, namely 4+. Interestingly, $^{101}\nu_Q$ at the Ru site in $LaRu_2Si_2$, in which La ions are supposed to be trivalent, is roughly twice larger than those for URu_2Si_2 and $ThRu_2Si_2$ [4]. A theoretical calculation attempting the reproduction of these ν_Q's will be crucial for microscopic understanding of electronic states in these compounds.

4. Summary

We have carried out ^{29}Si- and ^{99}Ru-NMR measurements on the non-f compound $ThRu_2Si_2$ as a nonmagnetic reference for URu_2Si_2. The results are the first detailed NMR data on $ThRu_2Si_2$ to our knowledge. Both Knight shifts measured by ^{29}Si- and ^{99}Ru-NMR are independent of temperature, in good agreement with the temperature dependence of χ. The temperature dependence of ν_Q at the Ru site is much weaker than the data on URu_2Si_2, suggesting electronic configuration in URu_2Si_2, including $5f$ electrons, considerably changes with temperature. From the close values of ν_Q in $ThRu_2Si_2$ and URu_2Si_2, it is likely that the U ions in URu_2Si_2 are in a nearly tetravalent state rather than in a trivalent state especially at high temperatures.

References

[1] Palstra T T M, Menovsky A A, van den Berg J, Dirkmaat A J, Kes P H, Nieuwenhuys G J and Mydosh J A 1985 *Phys. Rev. Lett.* **55** 2727
[2] Schlabitz W, Baumann J, Pollit B, Rauchschwalbe U, Mayer H M, Ahlheim U and Bredl C D 1986 *Z. Phys.* B **62** 171
[3] Maple M B, Chan J W, Dalichaouch Y, Kohara T, Rossel C, Torikachvili M S, McElfresh M W and Thompson J D 1986 *Phys. Rev. Lett.* **56** 185
[4] Emi N, Hamabata R, Nakayama D, Miki T, Koyama T, Ueda K, Mito T, Kohori Y, Matsumoto Y, Haga Y, Yamamoto E, Fisk Z and Tsujii N 2015 *J. Phys. Soc. Jpn.* **84** 063702
[5] Kohori Y, Matsuda K and Kohara T 1996 *J. Phys. Soc. Jpn.* **65** 1083
[6] Matsuda K, Kohori Y and Kohara T 1996 *J. Phys. Soc. Jpn.* **65** 679
[7] Matsumoto Y, Haga Y, Tateiwa N, Aoki H, Kimura N, Matsuda T D, Yamamoto E, Fisk Z and Yamagami H 2014 *JPS Conf. Proc.* **3** 011096
[8] Amitsuka H and Sakakibara T 1993 *J. Phys. Soc. Jpn.* **63** 736

IOP Conf. Series: Journal of Physics: Conf. Series **868** (2017) 012014 doi:10.1088/1742-6596/868/1/012014

[9] From the observation of ^{101}Ru-NQR signal at 4.2 K, $^{101}\nu_{\rm Q}$ is estimated to be 6.05 MHz at 4.2 K. Then $^{99}\nu_{\rm Q}$ is estimated to be 1.046 MHz using a relation of $^{101}Q/^{99}Q$=457/79 [11], where Q is nuclear quadrupole moment.
[10] Mito T, Hattori M, Motoyama G, Sakai Y, Koyama T, Ueda K, Kohara T, Yokoyama M and Amitsuka H 2013 *J. Phys. Soc. Jpn.* **82** 123704
[11] Pyykkö P 2008 *Mol. Phys.* **106** 1965

IOP Conf. Series: Journal of Physics: Conf. Series **868** (2017) 012015 doi:10.1088/1742-6596/868/1/012015

Magnetic properties of exchange-enhanced Pauli paramagnetic metals $AeCo_2P_2$ (Ae = Sr, Ba)

M. Imai[1], **C. Michioka**[1], **H. Ueda**[1], **A. Matsuo**[2], **K. Kindo**[2], **K. Yoshimura**[1,3]

[1]Department of Chemistry, Graduate School of Science, Kyoto University, Kyoto 606-8502, Japan
[2]Institute for Solid State Physics, The University of Tokyo, Kashiwanoha, Kashiwa, Chiba 277-8581, Japan
[3]Research Center for Low Temperature and Materials Sciences, Kyoto University, Kyoto 606-8501, Japan

E-mail: m.imai@kuchem.kyoto-u.ac.jp, kyhv@kuchem.kyoto-u.ac.jp

Abstract. We have synthesized single crystals of $AeCo_2P_2$ (Ae = Sr, Ba) and polycrystalline powder sample of $Sr_{1-x}Ba_xCo_2P_2$, and measured their magnetic properties. $SrCo_2P_2$ shows an itinerant-electron metamagnetic transition at H_c = 600 kOe and its temperature dependent magnetic susceptibility shows characteristic two anomalies at T_{max1} = 25 and T_{max2} = 115 K, suggesting it is in the vicinity of a ferromagnetic quantum critical point. While, $BaCo_2P_2$ does not show the metamagnetic transition with the applied field up to 600 kOe, and its magnetic susceptibility shows only one anomaly corresponding to T_{max1} at 33 K. In addition, the Weiss temperature of $Sr_{1-x}Ba_xCo_2P_2$, decreases to negative as x increases. These results indicate the system recede from the quantum critical point by the Ba substitution.

1. Introduction

Since the discovery of the iron pnictide superconductor $LaFeAsO_{1-x}F_x$ [1], layered transition metal pnictides and chalcogenides have been extensively studied. These iron based superconductivities are classified by the structure as 1111-type with ZrCuSiAs type structure [1], 122-type with $ThCr_2Si_2$ type structure [2], 111-type with anti-PbFCl structure [3], and so on, and they are based on transition metal layers in which the transition metals are coordinated by pnictogens and chalcogens.

The same structural compounds have been discovered in cobalt systems, e.g., ACo_2X_2 (A = Alkali metals, Alkaline earth metals, and rare earth metals; X = P, As, S, Se) [4, 5, 6, 7, 8, 9, 10, 11, 12], $LaCoPnO$ (Pn = P and As) [13, 14], and Sr_2ScO_3CoPn [15, 16]. In contrast to the iron system with antiferromagnetic interaction, these cobalt compounds have ferromagnetic interaction in their $CoPn$ layer, thus, they have been investigated as candidates of quasi-two dimensional itinerant-electron ferromagnets. For example, two dimensional ferromagnetic behavior in $LaCoPnO$ (Pn = P and As) is observed by magnetization and NMR measurements [13, 14], and highly magnetic anisotropic behavior is observed in single crystals of ACo_2Se_2 [11]. Furthermore, an itinerant-electron metamagnetic transition is discovered in $SrCo_2P_2$, suggesting it is in the vicinity of ferromagnetic quantum critical point [17]. In the Ca substituted system $Sr_{1-x}Ca_xCo_2P_2$, the metamagnetic transition field decreases as x increases to 0.5, and the antiferromagnetic ground state appears in the region of $x > 0.6$ [17]. In this system, magnetic properties are closely related to two types of crystal structure. One is the collapsed tetragonal (cT) structure with making interlayer P-P bond, and the other is the uncollapsed tetragonal (ucT) one. $Sr_{1-x}Ca_xCo_2P_2$ ($0 \leq x \leq 0.5$) with the ucT structure does not show any magnetic orderings and $Sr_{1-x}Ca_xCo_2P_2$ ($0.6 \leq x \leq 1$) with cT structure shows antiferromagnetic ordering [17, 18]. In this paper, taking up the magnetic properties of $Sr_{1-x}Ba_xCo_2P_2$ as a counterpart of $Sr_{1-x}Ca_xCo_2P_2$, we synthesized single crystals of $AeCo_2P_2$ (Ae = Sr, Ba) and polycrystalline samples of $Sr_{1-x}Ba_xCo_2P_2$ with the ucT structure, and measured their magnetic susceptibility and high magnetization process. We will show that temperature dependent susceptibility and magnetization curves of $BaCo_2P_2$ show an anomaly which is not the itinerant-electron metamagnetic transition. and that magnetic interactions of $Sr_{1-x}Ba_xCo_2P_2$ decreases by the Ba substitution.

2. Experimental method

Single crystalline and polycrystalline samples of $Sr_{1-x}Ba_xCo_2P_2$ were prepared from Sr(2N), Ba(2N), Co(3N) and P(red, 5N). The single crystals with $x = 0$ and 1 were obtained by a tin flux method [6]. Excess tin was removed by a dilute hydrochloric acid in the case of $SrCo_2P_2$. Because $BaCo_2P_2$ is easily soluble in dilute acids, excess tin was removed by centrifugation after heating to 673 K. Polycrystalline samples were synthesized as shown in Ref [17]. These samples were characterized by powder x-ray diffraction (XRD) using Cu Kα radiation. Polycrystalline samples and single crystals were placed on a glass plate and measured in Bragg-Brentano geometry.

The temperature dependent magnetizations of $Sr_{1-x}Ba_xCo_2P_2$ were measured by a Quantum Design MPMS-XL system at the Research Center for Low Temperature and Materials Sciences, Kyoto University. Magnetization curves beyond 700 kOe were measured by using an induction method with a multilayer pulsed magnet at the ultrahigh magnetic field laboratory of the Institute for Solid State Physics, the University of Tokyo.

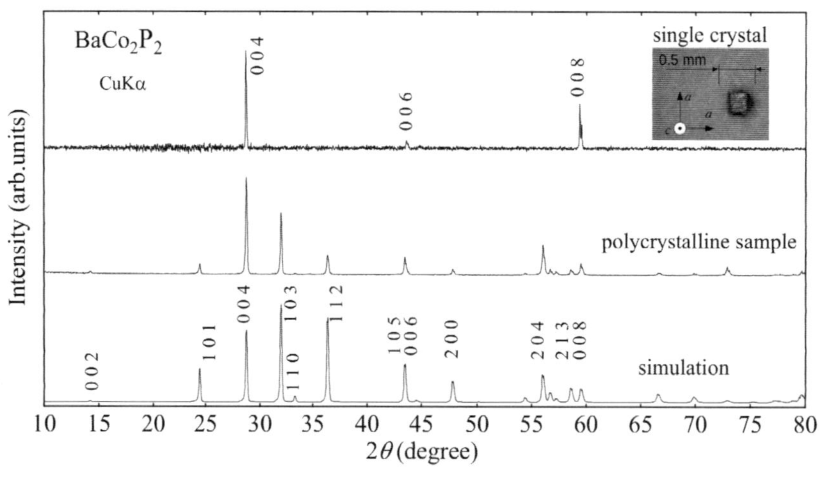

Figure 1. X-ray diffraction patterns of a single crystal, polycrystalline sample, and a simulated profile of $BaCo_2P_2$. The Miller indices of each peak are represented in the figure. The inset shows a photograph of a single crystal of $BaCo_2P_2$.

3. Results and discussions

Single crystals and polycrystalline sample of $SrCo_2P_2$ were obtained as described in previous work [4] and characterized by using XRD patterns. In the case of $BaCo_2P_2$, plate-like tetragonal crystals were obtained. The inset of Fig. 1 shows the photograph of a typical crystal of $BaCo_2P_2$. Figure 1 shows XRD patterns of a single crystal and a polycrystalline sample of $BaCo_2P_2$. All diffraction peaks are attributed to the $ThCr_2Si_2$ type structure with space group of $I4/mmm$. XRD patterns of $BaCo_2P_2$ were refined by the Rietveld method, using a computer program RIETAN-FP [19]. The lattice constants a = 3.8040(2) Å and c = 12.4186(5) Å are larger than those of $SrCo_2P_2$. The simulated pattern and Miller indices are shown in Fig. 1. The XRD pattern of the single crystal shows large 0 0 l Bragg peaks, and their positions are quite consistent with the lattice parameter c, suggesting successful synthesis of single crystals of $BaCo_2P_2$.

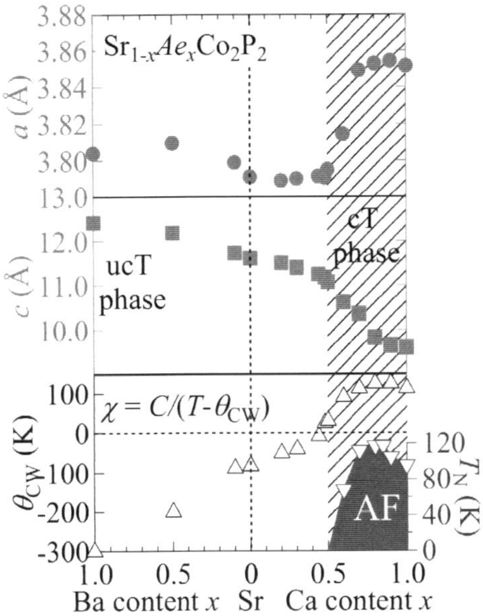

Figure 2. Lattice parameters and magnetic properties of $Sr_{1-x}Ae_xCo_2P_2$ (Ae = Ca, Ba). ●: a. ■: c. △: Weiss temperature. ▽: Néel temperature. The shaded region represents collapsed tetragonal (cT) region and the other does uncollapsed tetragonal (ucT) region. AF stands for the antiferromagnetic state.

Lattice parameters of $Sr_{1-x}Ba_xCo_2P_2$ (x = 0, 0.5) were determined from a refinement of the diffraction patterns by using the Le Bail method, using RIETAN-FP. These lattice parameters are shown in Fig. 2 with those of $Sr_{1-x}Ca_xCo_2P_2$ in Ref.[17]. As the ionic radius of alkaline earth cation Ae^{2+} decreases in the order of $Ba^{2+} > Sr^{2+} > Ca^{2+}$, the lattice parameter c corresponding to the inter-CoP-plane distance monotonically decreases. For the substituted system, the lattice parameter a is almost unchanged, and c decreases as the average radius of Sr^{2+} and the substituted Ba^{2+} or Ca^{2+} decreases in the ucT region. In this region, the inter-CoP-plane distance is deemed to be controlled by the cation substitutions and $BaCo_2P_2$ expects to have the weakest interlayer interaction in this system. The structural change from the ucT to cT structure occurs at the Ca content x = 0.5 with making of interlayer P-P bonds. Therefore, a increases rapidly and c decreases around x = 0.5, and magnetic ordering occurs in the cT region.

Figure 3 shows the temperature dependence of magnetic susceptibility $\chi(T)$ and its reciprocal of $AeCo_2P_2$ (Ae = Sr, Ba). As shown in Fig. 3(a), $\chi(T)$ of $SrCo_2P_2$ shows double maxima at 25 and 115 K and Curie-Weiss-like temperature dependence. The susceptibility with $H \parallel a$ is larger than that with $H \parallel a$, suggesting the easy axis is the c direction. The Weiss temperature θ_{CW} and the effective Bohr magneton number p_{eff} are −95.2 K and 1.59 for $H \parallel a$, and −90.0 K and 153

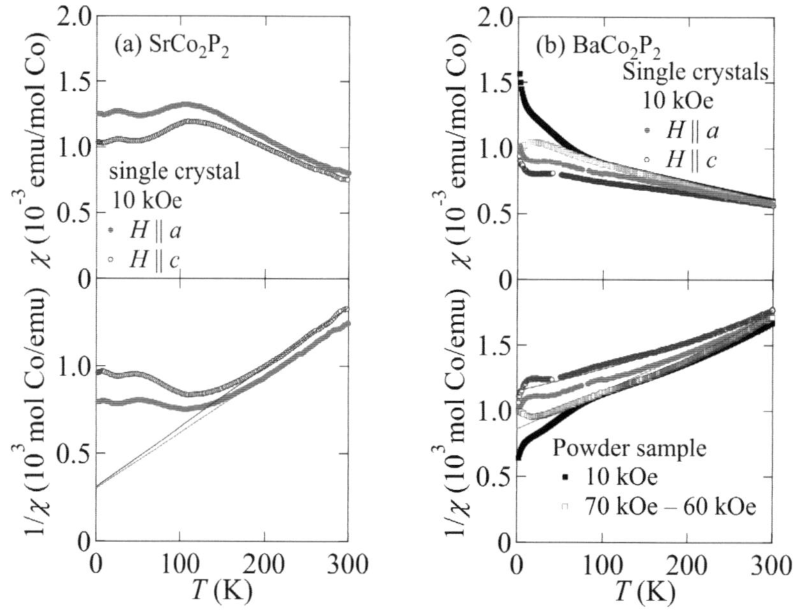

Figure 3. Temperature dependence of the magnetic susceptibility and its reciprocal. (a) The data for the single crystal of $SrCo_2P_2$. (b) The data for the single crystals and polycrystalline powder sample of $BaCo_2P_2$. Open and closed circles show the data for the single crystals with the applied field $H \parallel a$ and $H \parallel c$, respectively. Squares show the data for the polycrystalline powder sample. Open one shows the data with $H = 10$ kOe and closed one shows $\Delta M/\Delta H$ obtained from magnetization $M(H)$, here $\Delta M = M(70\text{ kOe}) - M(60\text{ kOe})$, $\Delta H = 70 - 60$ kOe.

for $H \parallel c$ from Curie-Weiss fit at between 200 to 300 K. In the case of itinerant magnetism with ferromagnetic interaction, the Curie-Weiss like temperature dependence originates in the temperature dependence of the amplitude of the local spin fluctuation, and the Weiss temperature is an indicator of the distance from the quantum critical point [20, 21, 22]. When the system approaches to the quantum critical point, the negative Weiss temperature increases to 0.

Figure 3(b) shows $\chi(T)$ of the polycrystalline powder sample and the single crystal of $BaCo_2P_2$. For both field directions, $\chi(T)$ of single crystal shows maximum at 33 K and Curie-Weiss-like temperature dependence at high temperature region. An upturn at low temperature seems to be ascribed to a magnetic impurity. For the powder sample, $\chi(T)$ shows larger upturn at low temperature and therefore does not show the maximum behavior. Thus, to estimate the intrinsic magnetic susceptibility, we calculated susceptibility as $\chi(T) = \Delta M(T)/\Delta H$, where $\Delta M(T)$ is difference between magnetization at 60 and 70 kOe, $\Delta H = 70 - 60$ kOe. It is revealed that the calculated susceptibility also shows the maximum behavior, suggesting this maximum behavior is intrinsic in $BaCo_2P_2$. From Curie-Weiss fit at between 200 to 300 K, θ_{CW} and p_{eff} are -300 K and 1.67 for $H \parallel a$, and -400 K and 1.81 for $H \parallel c$. The Weiss temperature is smaller than that of $SrCo_2P_2$, suggesting the system has weaker ferromagnetic interaction and recede from the quantum critical point.

Figure 4 shows $\chi(T)$ and magnetization process at 4.2 K for $Sr_{1-x}Ba_xCo_2P_2$. The maximum behavior was not observed at $x = 0.1$, 0.5 because of probably upturn at low temperature

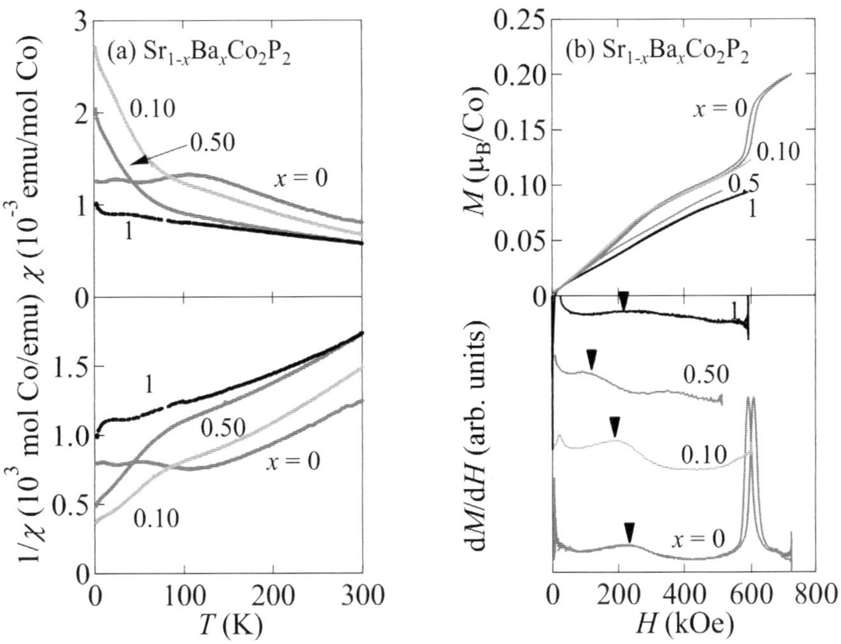

Figure 4. Magnetic properties of $Sr_{1-x}Ba_xCo_2P_2$. (a) Temperature dependence of the magnetic susceptibility and its reciprocal. (b) The magnetization M and the differential magnetization dM/dH curves measured at 4.2 K in pulsed high magnetic fields up to 700 kOe.

region. All compounds show Curie-Weiss like temperature dependence. As shown in the bottle panel of Fig. 2, θ_{CW} decreases continuously as x increases. Combined with the fact that the lattice parameter c increases with increasing x, the decrease of the magnetic interaction can be explained by the development of two dimensionality in the layered structure.

The magnetization curve of $SrCo_2P_2$ shows two anomalies at 240 and 600 kOe. The high-field anomaly corresponds to the itinerant-electron metamagnetic transition, which is the transition from the Pauli paramagnetic to itinerant ferromagnetic phases. On the other hand, the metamagnetic transition was not observed up to 600 kOe in Ba substituted system. For the result that the metamagnetic transition field decreases with the Ca substitution [17], it is natural to that the metamagnetic transition field increases beyond 600 kOe in the Ba substituted system. The low-field anomaly appears in all compounds of $Sr_{1-x}Ba_xCo_2P_2$, and it shows a little x dependence in comparison to metamagnetic transition field of $Sr_{1-x}Ca_xCo_2P_2$ [17], suggesting this anomaly is related to the lattice parameter a rather than c. In addition, this anomaly seems to be related to the small maximum behavior at low-temperature in $\chi(T)$. In this connection, the maximum behavior was not observed when P replaced by As [8, 23, 24, 10]. Thus, the maximum behavior and the low-field anomaly are peculiar behavior of the cobalt phosphide and would originate in their characteristic shape of the electric band structure near the Fermi level.

In this system, the mainly magnetic interaction causing the metamagnetic transition depends on the interlayer distance (proportional to c), and the low-field anomaly seems to depend on the intralayer Co distance (proportional to a) rather than the interlayer distance.

4. Conclusion

We have synthesized single crystals of $AeCo_2P_2$ (Ae = Sr, Ba) and polycrystalline samples of the solid solution system $Sr_{1-x}Ba_xCo_2P_2$, and investigated their magnetic properties. Magnetization curve of all compounds show a small anomaly at approximately 200 kOe. The anomaly has a little x dependence in comparison to metamagnetic transition and seems to originate in the characteristic shape of the electric band structure near the Fermi level, depending on the intralayer structure rather than the interlayer structure. While, Ba substituted compounds do not show the metamagnetic transition below 600 kOe and their Weiss temperature systematically decreases with increasing x. These results indicate the system moves away from the quantum critical point with increase of the interlayer distance by Ba substitution.

References

[1] Kamihara Y, Watanabe T, Hirano M and Hosono H 2008 *Journal of the American Chemical Society* **130** 3296–3297 pMID: 18293989 (*Preprint* `http://dx.doi.org/10.1021/ja800073m`) URL `http://dx.doi.org/10.1021/ja800073m`
[2] Rotter M, Tegel M and Johrendt D 2008 *Phys. Rev. Lett.* **101**(10) 107006 URL `http://link.aps.org/doi/10.1103/PhysRevLett.101.107006`
[3] Wang X, Liu Q, Lv Y, Gao W, Yang L, Yu R, Li F and Jin C 2008 *Solid State Communications* **148** 538 – 540 ISSN 0038-1098 URL `http://www.sciencedirect.com/science/article/pii/S0038109808005851`
[4] Mörsen E, Mosel B, Müller-Warmuth W, Reehuis M and Jeitschko W 1988 *J. Phys. Chem. Solids* **49** 785–795 URL `http://www.sciencedirect.com/science/article/pii/0022369788900303`
[5] Huan G and Greenblatt M 1989 *Journal of the Less Common Metals* **156** 247 – 257 ISSN 0022-5088 URL `http://www.sciencedirect.com/science/article/pii/0022508889904232`
[6] Reehuis M, Ritter C, Ballou R and Jeitschko W 1994 *J. Magn. Magn. Mater.* **138** 85–93 URL `http://www.sciencedirect.com/science/article/pii/0304885394904022`
[7] Reehuis M, Jeitschko W, Kotzyba G, Zimmer B and Hu X 1998 *J. Alloys Compd.* **266** 54–60 URL `http://www.sciencedirect.com/science/article/pii/S0925838897004866`
[8] Sefat A, Singh D J, Jin R, McGuire M A, Sales B C and Mandrus D 2009 *Phys. Rev. B* **79** 024512
[9] Cheng B, Hu B, Yuan R, Dong T, Fang A, Chen Z, Xu G, Shi Y, Zheng P, Luo J *et al.* 2012 *Phys. Rev. B* **85** 144426
[10] Pandey A, Quirinale D G, Jayasekara W, Sapkota A, Kim M G, Dhaka R S, Lee Y, Heitmann T W, Stephens P W, Ogloblichev V, Kreyssig A, McQueeney R J, Goldman A I, Kaminski A, Harmon B N, Furukawa Y and Johnston D C 2013 *Phys. Rev. B* **88**(1) 014526 URL `http://link.aps.org/doi/10.1103/PhysRevB.88.014526`
[11] Yang J, Chen B, Wang H, Mao Q, Imai M, Yoshimura K and Fang M 2013 *Phys. Rev. B* **88**(6) 064406 URL `http://link.aps.org/doi/10.1103/PhysRevB.88.064406`
[12] Thompson C M, Tan X, Kovnir K, Garlea V O, Gippius A A, Yaroslavtsev A A, Menushenkov A P, Chernikov R V, Büttgen N, Krätschmer W, Zubavichus Y V and Shatruk M 2014 *Chemistry of Materials* **26** 3825–3837 (*Preprint* `http://dx.doi.org/10.1021/cm501522v`) URL `http://dx.doi.org/10.1021/cm501522v`
[13] Yanagi H, Kawamura R, Kamiya T, Kamihara Y, Hirano M, Nakamura T, Osawa H and Hosono H 2008 *Phys. Rev. B* **77**(22) 224431 URL `http://link.aps.org/doi/10.1103/PhysRevB.77.224431`
[14] Ohta H and Kazuyoshi Y 2009 *Phys. Rev. B* **79**(18) 184407 URL `http://link.aps.org/doi/10.1103/PhysRevB.79.184407`
[15] Okada S, Kamihara Y, Ohkubo N, Ban S, Matoba M and Atou T 2012 *Journal of Physics: Conference Series* **400** 022091 URL `http://stacks.iop.org/1742-6596/400/i=2/a=022091`
[16] Ohta H, Noguchi D, Nabetani K and Katori H A 2013 *Phys. Rev. B* **88**(9) 094441 URL `http://link.aps.org/doi/10.1103/PhysRevB.88.094441`
[17] Imai M, Michioka C, Ohta H, Matsuo A, Kindo K, Ueda H and Yoshimura K 2014 *Phys. Rev. B* **90** 014407
[18] Jia S, Williams A J, Stephens P W and Cava R J 2009 *Phys. Rev. B* **80**(16) 165107 URL `http://link.aps.org/doi/10.1103/PhysRevB.80.165107`
[19] Izumi F and Momma K 2007 *Solid State Phenomena* **130** 15–20 URL `http://www.scientific.net/SSP.130.15`
[20] Moriya T 1985 *Spin fluctuations in itinerant electron magnetism* vol 56 (Berlin Heidelberg: Springer-Verlag)
[21] Takahashi Y 1986 *J. Phys. Soc. Jpn.* **55** 3553–3573 (*Preprint* `http://dx.doi.org/10.1143/JPSJ.55.3553`) URL `http://dx.doi.org/10.1143/JPSJ.55.3553`
[22] Takahashi Y 2013 *Spin Fluctuation Theory of Itinerant Electron Magnetism* (Berlin Heidelberg: Springer-Verlag)

IOP Conf. Series: Journal of Physics: Conf. Series **868** (2017) 012015 doi:10.1088/1742-6596/868/1/012015

[23] Ahilan K, Imai T, Sefat A and Ning F 2014 *Phys. Rev. B* **90** 014520
[24] Wiecki P, Ogloblichev V, Pandey A, Johnston D and Furukawa Y 2015 *Phys. Rev. B* **91** 220406

Electron correlation in Pauli paramagnetic Cr_2AlC, Cr_2GaC and Cr_2GeC

Zhongsheng Liu, Kenta Takao, Takeshi Waki, Yoshikazu Tabata, and Hiroyuki Nakamura

Department of Materials Science and Engineering, Kyoto University, Kyoto 606-8501, Japan

E-mail: `liuzhongsheng04@gmail.com`

Abstract. Cr_2AlC, Cr_2GaC, and Cr_2GeC are classical MAX phase compounds, for which successful synthesis of bulk equilibrium phases has been reported in an early stage. Although it has been established that they are Pauli paramagnetic down to the lowest temperature, the extent of the exchange enhancement depends on the A element (A = Al, Ga, and Ge). We discuss the nature of electron correlation by analyzing low-temperature resistivity, specific heat, and susceptibility in terms of the Kadowaki-Woods and Wilson ratios.

1. Introduction

The $M_{n+1}AX_n$ (space group $P6_3/mmc$, abbreviated as MAX) phases are layered, hexagonal, carbides and nitrides, where M is an early transition metal, A is an A-group (mostly groups IIIA and IVA) element, and X is C and/or N, and n = 1, 2, 3, ... [1]. MAX phases have attracted considerable attention recently due to their unique material properties combining the characteristics of metals and ceramics and the presence of a large number of isostructural compounds. MAX phases are excellent electrical and thermal conductors.

When M is the $3d$ transition metal element, only materials consisting of M heavier than Cr are known to exist as equilibrium phases. The pure Cr_2AC (A = Al, Ga, and Ge) compounds show Pauli paramagnetic behaviors down to the lowest temperature [2]. The magnetism is more exchange-enhanced in the order of Cr_2GeC, Cr_2GaC, and Cr_2AlC. On the other hand in Cr_2GaN, which is only one Cr-based nitride ever known, a magnetic phase transition, to the spin-density-wave (SDW) state, has been found at ~170 K [3]. The origin of the SDW transition is interpreted in terms of possible Fermi-surface nesting in the two-dimensional-like electronic structure.

The purpose of this report is to discuss the nature of exchange-enhanced Pauli paramagnetism, i.e., electron correlation, in the Cr-based MAX phases, Cr_2AC (A = Al, Ga, and Ge). NMR result of Cr_2AlC is also included to prove unambiguously no long-range magnetic order in the compound.

2. Experimental procedures

Polycrystalline samples of Cr_2AC (A = Al, Ga, and Ge) were synthesized via solid-state reactions in evacuated quartz tubes. The detailed procedures have been described in elsewhere [2, 3]. The temperature dependence of the resistivity was measured for pellets sintered from powder by the conventional four-probe method at 5–300 K. Nuclear magnetic resonance (NMR) measurements

IOP Conf. Series: Journal of Physics: Conf. Series **868** (2017) 012016 doi:10.1088/1742-6596/868/1/012016

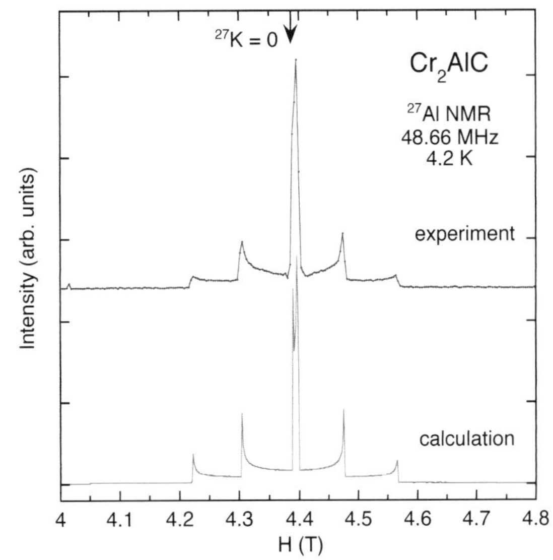

Figure 1. The field-swept NMR spectrum measured for Cr_2AlC at a frequency of 48.66 MHz and at 4.2 K. The simulated paramagnetic powder-pattern spectrum is also presented for comparison.

were performed for Cr_2AlC by a phase-coherent-type pulse spectrometer (Thamway, PROT II 2101MR). For the measurements, fine particles of the sample are soaked in paraffin to fix crystal axes at random. Spin-echo intensity was integrated and recorded as a function of external field to obtain the spectrum at 4.2 K. The nuclear spin-lattice relaxation time, T_1, was measured by the saturation-recovery method.

3. Results

The ^{27}Al NMR spectrum of Cr_2AlC down to 77 K has already been reported by Lue et al. [5]. The spectrum shows a typical paramagnetic pattern. It is already no doubt that Cr_2AlC has no long-range magnetic ordering. Here just for confirmation, we extended the measurement down to 4.2 K. Figure 1 shows a field-swept spectrum of ^{27}Al NMR measured at 4.2 K, which shows the typical paramagnetic powder pattern. This line shape is interpreted as follows. Nuclear parameters of ^{27}Al nucleus are natural abundance 100%, nuclear spin $I = 5/2$, the gyromagnetic ratio, $^{27}\ /2\pi = 11.093$ MHz/T, and the nuclear quadrupole moment $^{27}Q = 0.15 \quad 10 \ ^{28}$ m ^{2}. The Al site in the Cr_2AlC-type structure is crystallographically unique ($2d$) and axially symmetric ($6m2$), resulting in the nuclear quadrupole interaction with zero asymmetry parameter ($= 0$). In the strong-field condition, we expect to observe two pairs of first-order quadrupolar satellites together with a center line. In the bottom of Fig. 1, the computer-generated powder pattern assuming the quadrupole frequency of 1.9 MHz and the absence of the internal field is presented. Here we neglected a possible hyperfine shift, for simplicity; the values of the Knight shift estimated from the experimental spectrum are small in accordance with those estimated at 77 K in Ref. [5] ($K_{iso} \sim -0.2\%$, $K_{aniso} \sim 0.01\%$). In this condition, the center line has a fine structure due to the second-order quadrupolar interaction. The calculated pattern agrees well with the experimental spectrum, proving unambiguously that Cr_2AlC has no long-range magnetic order at the temperature.

The ^{27}Al nuclear spin-lattice relaxation time, T_1, was measured with use of the central transition of the spectrum. The multi-exponential recovery of the nuclear magnetization was fitted to the function expected for the $I = 5/2$ nucleus [4]. The temperature dependence of the

Table 1. Lattice parameters a and c, lattice volume V, the electronic specific heat , the density of states at the Fermi level $N(E_F)$ estimated from band calculations, the zero-temperature susceptibility $\chi(0)$, and the coefficient of the T^2 term in the low-temperature resistivity A for Cr_2AC (A = Al, Ga, and Ge).

Compound	a (Å)	c (Å)	V (Å^3)	γ (mJ mol ^{1}K ^{2})	$N(E_F)$ ((eV unit cell) ^{1})	$\chi(0)$ (10 ^{4} emu/f.u.-mol)	A (10 $^{3}\mu\Omega$ cm K ^{2})
Cr_2AlC	2.862	12.832	91.023	17.15 [6] 15.12 [7]	14.6 [6] 12.9 [7]	2.75	1.0
Cr_2GaC	2.898	12.651	92.011	20.25 [8]	17.2 [8]	3.59	6.5
Cr_2GeC	2.961	12.122	92.038	26.12 [9] 25.00 [10]	22.0 [9] 21.0 [10]	5.50	5.6

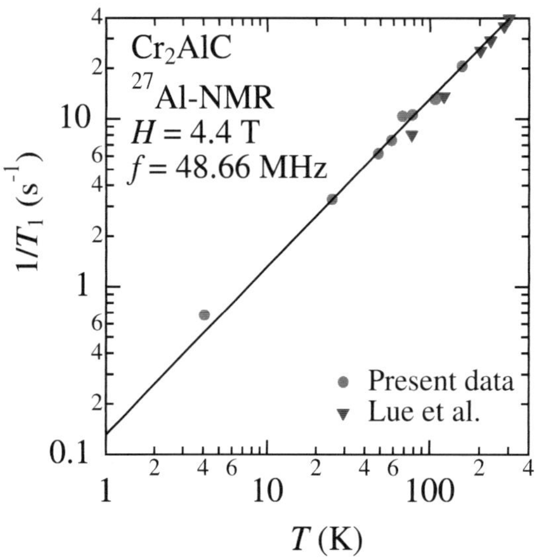

Figure 2. The temperature dependence of the ^{27}Al nuclear spin-lattice relaxation rate, $1/T_1$, measured for Cr_2AlC. The straight line indicates the Korringa relaxation. The data points indicated by triangles are cited from Ref. [5].

relaxation rate, $1/T_1$, down to 4.2 K is shown in Fig. 2. Our data are in good accordance with those above 77 K reported by Lue et al. [5]. The result of $1/T_1$ is in proportion to temperature down to the lowest temperature, namely, follows the Korringa law in all the temperature range, indicating that the relaxation is dominated by conduction electrons. This result also demonstrates that Cr_2AlC is a Pauli paramagnetic metal down to the lowest temperature.

We have also measured the Ga-NMR for Cr_2GaC, which was reported in Ref. [3]. Similarly to Cr_2AlC, Cr_2GaC shows a well resolved nonmagnetic spectrum, indicating that Cr_2GaC has no long-range magnetic order either.

Temperature dependencies of the resistivity for Cr_2AC (A = Al, Ga, and Ge) are shown in Fig. 3; the data for Cr_2GaC have already been published in Ref. [3]. They show monotonic and metallic temperature dependencies in agreement with previous data [8, 10, 11]. Here we make no mention of the origin of the residual resistivity, since the resistivity was measured for pellets sintered from powder. At low temperatures, the resistivity obeys the quadratic temperature dependance as $\rho = AT^2 + \rho_0$, where A and ρ_0 are constants, indicating the presence of appreciable

IOP Conf. Series: Journal of Physics: Conf. Series **868** (2017) 012016 doi:10.1088/1742-6596/868/1/012016

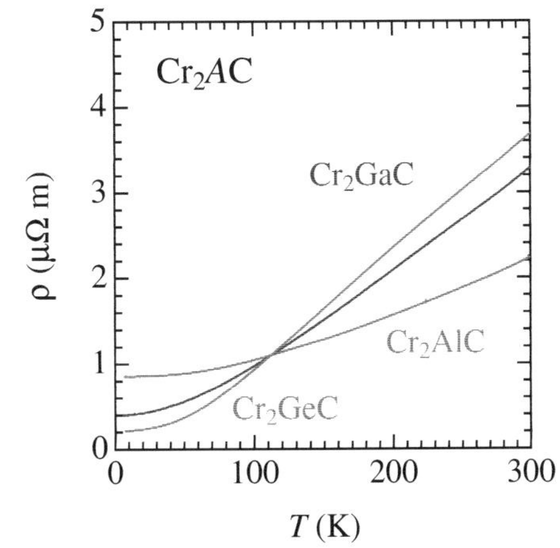

Figure 3. Temperature dependencies of the resistivity for Cr_2AC (A = Al, Ga, and Ge).

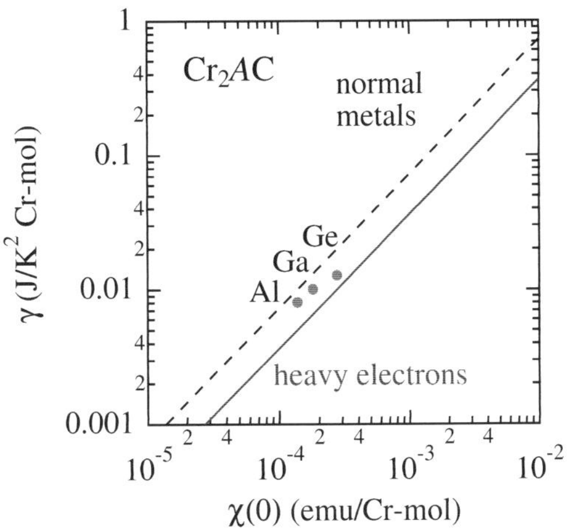

Figure 4. The relation between and $\chi(0)$. for Cr_2AlC is the average of values reported in Refs. [6] and [7]. for Cr_2GaC was cited from Ref. [8]. for Cr_2GeC is the average of values given in Refs. [9] and [10]. The solid and broken lines indicate the Wilson ratios for heavy electron systems and normal metals, respectively.

electron correlation.

4. Discussion

Cr_2AC (A = Al, Ga, and Ge) show typical Pauli paramagnetic behavior without spontaneous magnetization in all temperature range. No long-range magnetic order has been confirmed microscopically in Cr_2AlC and Cr_2GaC from NMR experiments. The absence of internal field unambiguously excludes the possibility of antiferromagnetism, to say nothing of ferro- and ferrimagnetism. The result of neutron diffraction of bulk Cr_2AlC, which does not show any additional re ections at low temperature [12], is thoroughly consistent with no long-range order. The absence of a phase transition for Cr_2GeC in the temperature range of 3–1500 K has already been confirmed in specific heat measurements [9].

Results of band structure calculations tell us that Cr_2AC are typical metals in the sense that

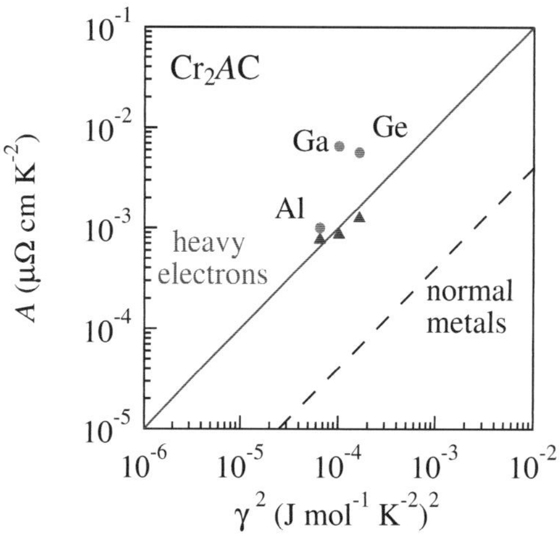

Figure 5. The coefficient of the quadratic term of the low-temperature resistivity, A, plotted against ^{2} (solid circles). The values estimated from the resistivity given in literatures [8, 10, 11] are also included (solid triangles). See the caption of Fig. 4 for . The solid and broken lines indicate the Kadowaki-Woods ratios for heavy electron systems and normal metals, respectively.

Cr-3d bands dominate the density of states at the Fermi level, $N(E_F)$. The systematic increase in χ from Cr_2AlC to Cr_2GeC is understood as the increase in $N(E_F)$, associated with the decrease of the 3d band width. Furthermore, the increase of the dimensionality expected from the decrease in the interlayer distance may also give rise to magnetic enhancement. The comparison of the electronic specific heat coefficient, , (see Table 1) leads to the same discussion. The susceptibility at 0 K, $\chi(0)$, which was estimated from the extrapolation of the high-temperature susceptibility [3] to 0 K, is plotted against in Fig. 4. The susceptibility and are both proportional to $N(E_F)$, as $= \pi^2 k_B^2 N(E_F)/3$ and $\chi = \ {}_B^2 N(E_F)$, respectively. The theoretical values of $\chi(0)/$, the so-called Wilson ratio, are also shown by the solid and broken lines for heavy electron systems and normal metals, respectively. The experimental points for Cr_2AC (A = Al, Ga, and Ge) deviate from the broken line of normal metals toward the solid line of heavy electron systems, indicating that Cr_2AC are not simple metals but more or less exchange-enhanced.

The resistivity were fit by the equation of $\rho = AT^2 + \rho_0$ in the temperature range of 5–35 K to estimate A values, which are plotted in Fig. 5 against ^{2}. The Kadowaki-Woods ratios, $A/\ ^2$, expected for heavy electron systems and normal metals are also shown in the same figure by the solid and broken lines, respectively. The points for Cr_2AC (A = Al, Ga, and Ge) are close to the solid line, indicating that Cr_2AC belong to the correlated electron system. The mass enhancement of Cr_2GeC compared with bare $N(E_F)$ by a factor of 3, pointed out in Ref. [13], also looks reasonable.

In summary, in the Pauli paramagnetic Cr-based MAX phases, electron correlation plays an appreciable role, and the correlation is more enhanced in the order of Cr_2GeC, Cr_2GaC and Cr_2AlC.

References

[1] Barsoum M W 2013 *MAX Phases: Properties of machinable ternary carbides and nitrides* (Weinheim: Wiley-VCH)
[2] Liu Z, Waki T, Tabata Y and Nakamura H 2014 *Phys. Rev.* B **89** 054435
[3] Liu Z, Waki T, Tabata Y, Yuge K, Nakamura H and Watanabe I 2013 *Phys. Rev.* B **88** 134401
[4] Narath A 1967 *Phys. Rev.* **162** 320

IOP Conf. Series: Journal of Physics: Conf. Series **868** (2017) 012016 doi:10.1088/1742-6596/868/1/012016

[5] Lue C S, Lin J Y and Xie B X 2006 *Phys. Rev.* B **73** 035125
[6] Drulis M K, Drulis H, Gupta S, Barsoum M W and El-Raghy T 2006 *J. Appl. Phys.* **99** 093502
[7] Lo and S E, Hettinger J D, Harrell K, Finkel P, Gupta S, Barsoum M W and Hug G 2004 *Appl. Phys. Lett.* **84** 508
[8] Lin S, Tong P, Wang B S, Huang Y N, Lu W J, Shao D F, Zhao B C, Song W H and Sun Y P 2013 *J. Appl. Phys.* **113** 053502
[9] Drulis M K, Drulis H, Hackemer A E, Leaffer O, Spanier J, Amini S, Barsoum M W, Guilbert T and El-Raghy T *J. Appl. Phys.* **104** 023526
[10] Barsoum M W, Scabarozi T H, Amini S, Hettinger J D and Lo and S E 2011 *J. Am. Ceram. Soc.* **94** 4123
[11] Hettinger J D, Lo and S E, Finkel P, Meehan T, Palma J, Harrell K, Gupta S, Ganguly A, El-Raghy T and Barsoum M W 2005 *Phys. Rev.* B **72** 115120
[12] Jaouen M, Chartier P, Cabioc'h T, Mauchamp V, Andre G and Viret M 2013 *J. Am. Ceram. Soc.* **96** 3872
[13] Eklund P, Bugnet M, Mauchamp V, Dubois S, Tromas C, Jensen J, Piraux L, Gence L, Jaouen M and Cabioc'h T 2011 *Phys. Rev.* B **84** 075424

Spin wave dispersion just above the magnetic order-order transition in the metallic antiferromagnet Mn_3Pt

Soshi Ibuka[1], Tetsuya Yokoo[1], Shinichi Itoh[1], Kazuya Kamazawa[2], Mitsutaka Nakamura[3] and Motoharu Imai[4]

[1] Institute of Materials Structure Science, High Energy Accelerator Research Organization, Tokai, Ibaraki 319-1106, Japan

[2] Center for Neutron Science and Technology, Comprehensive Research Organization for Science and Society, Tokai, Ibaraki 319-1106, Japan

[3] Materials and Life Science Division, J-PARC Center, Japan Atomic Energy Agency, Tokai, Ibaraki 319-1195, Japan

[4] National Institute for Materials Science, Tsukuba, Ibaraki 305-0047, Japan

E-mail: `ibuka@post.j-parc.jp`

Abstract. Spin wave dispersion of the metallic antiferromagnet Mn_3Pt was investigated just above the order-order transition temperature by using the inelastic neutron scattering technique. The spin wave dispersion at $T = 400$ K along [100], [110] and [111] directions was isotropic within the measurement accuracy. The dispersion was described by $(\hbar\omega)^2 = c^2q^2 + \Delta^2$ with $c = 190$ meVÅ and $\Delta = 3.3$ meV. Compared with the previous reports, the result shows that large reduction of the stiffness constant with increasing temperature. This feature is similar to that in $FePt_3$, and will be an indication of the itinerancy of the magnetic moments.

1. Introduction

The recent discovery of iron-based superconductors [1] has renewed interest in a basic understanding of the dynamical magnetic properties of metallic antiferromagnets. In metallic antiferromagnets with $4d$ or $5d$ electrons like $FePt_3$ [2], the spin dynamics are expected to show the intermediate character between itinerant magnets, such as Cr and γ-FeMn, and localized spin magnets, such as Heusler alloys, because of the high polarizability of $4d$ and $5d$ electrons [3]. The understanding of the spin dynamics in these magnets is still incomplete. Thus, the aim of this study is to examine the itinerancy and the locality of the magnetic moments in one of these alloys, Mn_3Pt.

Mn_3Pt is a Cu_3Au-type metallic antiferromagnet with the transition temperature T_N = 475 K [4, 5]. It shows magnetic order-order transition at $T_t = 400$ K. In the high-temperature ordered phase, which is known as F-phase, two third of Mn moments are ordered collinearly and one third of Mn moments are not ordered. The magnetic structure was shown in Fig. 1(a). A slight tetragonal distortion $c/a = 0.9985$ occurs at T_N [6], reflecting the magnetic symmetry of the partial disordered state. In the low temperature phase, which is called D-phase, Mn moments form a triangular spin structure with the moments lying in the 111 plane. The F-

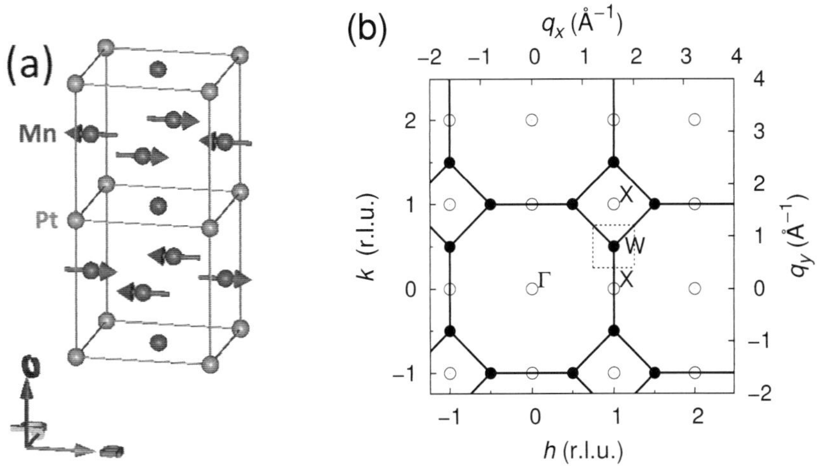

Figure 1. (a) Magnetic structure in the F-phase of Mn_3Pt. Pt ions do not have a magnetic moment. Two-third of Mn moments ordered with $\boldsymbol{Q}_{\rm F} = (1, 0, 0.5)$. (b) Brillouin zone of Mn_3Pt in $hk0$ plane. The open and close circles stand for the nuclear and magnetic elastic peak positions, respectively. The magnetic Brillouin zone was shown by the dotted line around a W point.

to D-phase transition leads to dramatic decreases in Mn moment from 3.4 to 2.2$\mu_{\rm B}$/Mn and in lattice volume by 2.25% [4, 7].

The stability of the magnetic structures was described by localized models. Krén et al have explained [5] the magnetic transition within the molecular-field approximation with the nearest-neighbor J_1 and the next-nearest-neighbor J_2 Mn-Mn exchange interactions, and showed that the exchange interactions are sensitive to the interatomic distances by studying the composition dependence of the magnetic structure. Ricodeau has introduced [8] the volume independent nearest-neighbor interaction J, and pointed out the importance of the taking into account of the entropy of the disordered Mn moments for stabilizing the F-phase. On the other hand, the spin wave in the D-phase has clear itinerant characters. The spin wave stiffness is as large as 240 meVÅ at $T = 20$ K [9]. The steep spin dispersion is the feature of the itinerant magnet. In addition, large reduction of the stiffness constant with increasing temperature, similar to that in $FePt_3$, was observed in the D-phase [9, 10]. In the F-phase, the stiffness constant is reduced to 130 meVÅ at 419 K, and 120 meVÅ at 459 K [11, 10]. With further increasing temperature, two-dimensional special short-range correlations developed [6, 12, 13]. The itinerant feature of the spin wave is seemingly weaker in the F-phase than in the D-phase. An investigation of the variation of the spin dynamics for a different temperature in the F-phase would give useful information to understand the spin dynamics of Mn_3Pt. Therefore, in this study, we performed inelastic neutron scattering experiments at $T = 400$ K just above $T_{\rm t}$.

2. Experiments

A single crystal of Mn_3Pt was synthesized by the Bridgman method. Mn chips and Pt wire were weighed with a molar ratio of Mn:Pt = 3.15:1. The purity was 99.99% for Mn and 99.99% for Pt. First, they were melted with using arc furnace under an argon gas atmosphere. The resulting alloy was placed in an alumina crucible, sealed in an electrically-fused quartz tube under an argon gas atmosphere, set in a vertical Bridgman furnace, and heated up to 1513 K. Next, it was slowly cooled down to 1413 K in 6 hours by moving it down to the bottom of the furnace.

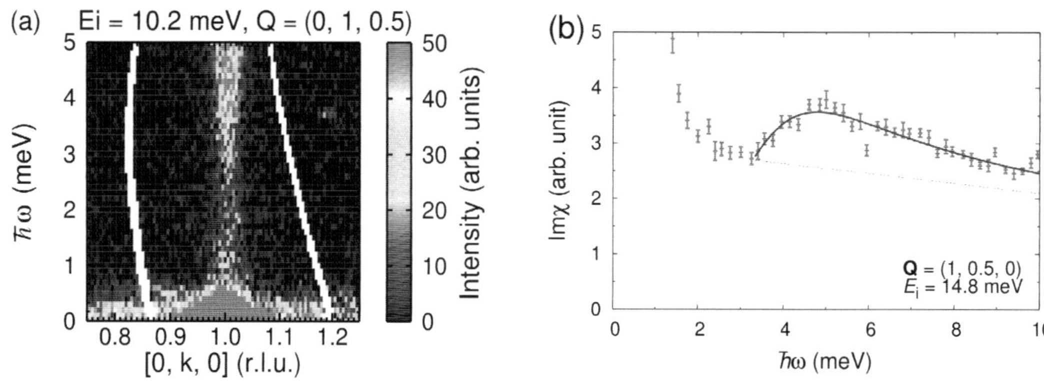

Figure 2. (a) Intensity map with the excitation energy $\hbar\omega$ and k around $\boldsymbol{Q} = (0, 1, 0.5)$ with $E_\mathrm{i} = 10.2$ meV. (b) $\hbar\omega$ dependence of $\mathrm{Im}\chi(\boldsymbol{Q}, \hbar\omega)$ at $\boldsymbol{Q} = (1, 0.5, 0)$ with $E_\mathrm{i} = 14.8$ meV. The solid line denotes the fit with a Lorentz function. The linear dotted line stands for background.

The temperature gradient of the furnace was 12.5 K/cm. Then, the sample was annealed at 1173 K for 90 h. The mass of the grown single crystal was 20 g. The chemical composition was determined by energy-dispersive X-ray spectroscopy with a scanning electron microscope (TM3030Plus and Quantax70, Hitachi High-Technologies). The chemical composition was Mn:Pt = 75.5(22):24.5(9). No impurity phase was found. The crystal orientation was determined by the x-ray Laue method with using YXLON MG452 (450 kV, 5 mA) and was aligned in $hk0$ zone. The Brillouin zone in $hk0$ plane was shown in Fig. 1(b). Inelastic neutron scattering experiments were performed by using the chopper spectrometer 4SEASONS installed at J-PARC/MLF, Japan. DAVE/MSlice [14] was used for analyzing the data. Incident neutron energies were set to E_i = 10.2, 14.8, 23.5, 42.9 and 102 meV with the energy resolution at the elastic position being 0.45, 0.7, 1.2, 2.5 and 8 meV, respectively. The crystal was sealed in an aluminum can under a ^{4}He gas atmosphere and then set in a closed cycle ^{4}He cryostat. Measurements were performed at $T = 400$ K. We confirmed that magnetic elastic peaks emerged at the W point, indicating that at least part of the sample was in F-phase.

3. Results

First, an energy gap of the spin wave Δ was investigated by using low-energy incident neutrons. Figure 2(a) shows the neutron scattering intensity map with the excitation energy $\hbar\omega$ and k around a W point. A spin excitation existed at $k = 1$. Figure 2(b) presents $\hbar\omega$ dependence of the imaginary part of the dynamic magnetic susceptibility $\mathrm{Im}\chi(\boldsymbol{Q}, \hbar\omega)$ at $\boldsymbol{Q} = (1, 0.5, 0)$. Δ was estimated to be 3.3(1) meV. In addition, a faint spin excitation was found even below the gap in Fig. 2(a), which is possibly another acoustic spin wave mode. We focus the former mode in the following sentences of this section, because the intensity of the latter mode was relatively weak.

Second, the spin wave dispersion was investigated by using high-energy incident neutrons. Figure 3(a) shows the typical intensity map with h and k at $\hbar\omega = 20$ meV. The spin wave dispersion along [100], [110] and [111] direction was estimated and shown in Fig. 3(b). The dispersion was isotropic within the measurement accuracy. A fitting with the equation, $(\hbar\omega)^2 = c^2q^2 + \Delta^2$, yields the spin wave stiffness $c = 190(10)$ meVÅ, where q is the wave number away from a W point, and Δ is an energy gap fixed to 3.3 meV. At 419 K, c and Δ were reported to be 130 meVÅ and 3.5 meV by the previous report [10]. The result indicates that

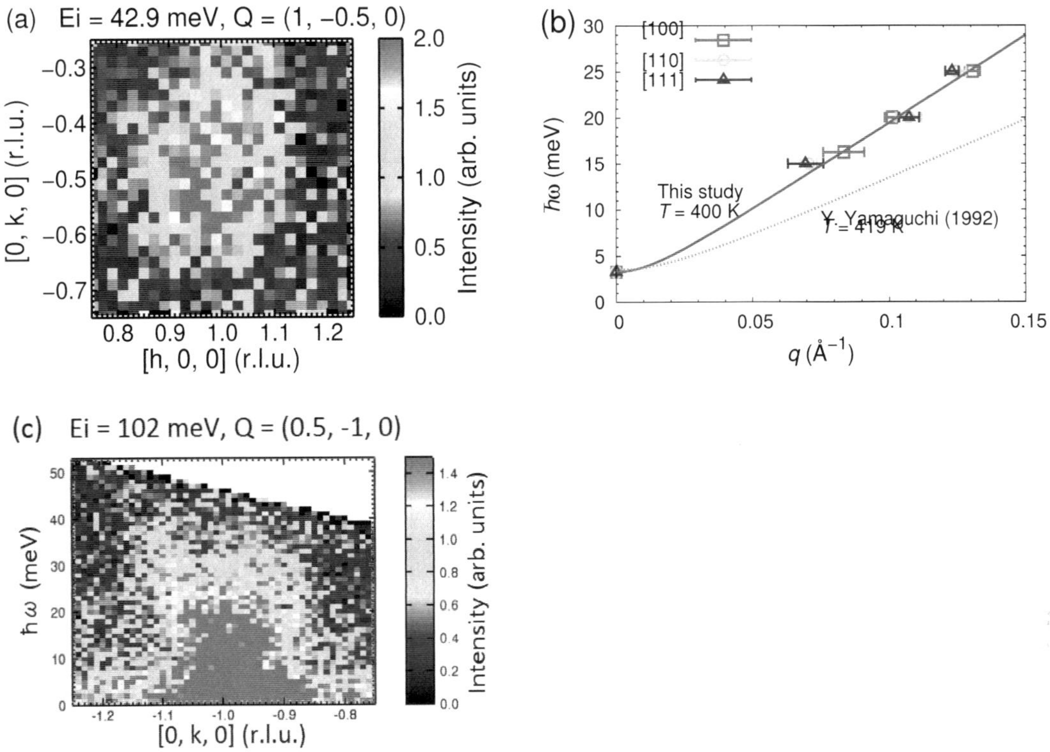

Figure 3. (a) Intensity map with h and k around $\boldsymbol{Q} = (1, -0.5, 0)$ at $\hbar\omega = 20$ meV with $E_{\rm i} = 42.9$ meV. (b) The spin wave dispersion along [100], [110] and [111] direction. The solid line shows the fit with the equation, $(\hbar\omega)^2 = c^2q^2 + \Delta^2$. The dotted line denotes the dispersion curve at $T = 419$ K reported in [10]. (c) Intensity map with $\hbar\omega$ and k around $\boldsymbol{Q} = (0.5, -1, 0)$ with $E_{\rm i} = 102$ meV.

the decreasing temperature from 419 K $= 0.88T_{\rm N}$ to 400 K $= 0.84T_{\rm N}$ leads to enhance c as much as 50%, while it has little influence on Δ. Figure 3(c) provides the intensity map with $\hbar\omega$ and k in high-energy region. The dispersion does not reach the zone boundary at $k = -1.25$ and -0.75 r.l.u. even at $\hbar\omega = 40$ meV. It was reported that the dispersion reaches the zone boundary at about 30 meV at 459 K $= 0.97T_{\rm N}$ [11]. The result suggests that the spin wave energy at the zone boundary increases with decreasing temperature, which is consistent with the increase of c with decreasing temperature.

4. Discussion

A large reduction in c with increasing temperature was observed even in the F-phase, which will be an indication of the itinerancy of the magnetic moments. This character is similar to that observed in another $5d$ alloys, $FePt_3$ [2]. On the other hand, if we discuss it within the molecular-field approximation, the variation in c with keeping Δ constant indicates a variation in the exchange interaction between the spins. Suppose that Δ originates from single-ion anisotropy, the unchanged Δ indicates that both the anisotropy constant and the moment size remain constant. The previous study of the temperature dependence of the moment size [7] supports the result; 3.4 and $3.2\mu_{\rm B}$ at $T = 400$ and 419 K, respectively. c depends on moment size and exchange interaction. Therefore, a variation in c result in a variation in the exchange interaction.

The previous inelastic neutron scattering study [10] found that the exchange interaction is different between the F- and D- phase, and pointed out that the difference originated in the band structure. Our result shows that the exchange interaction varies even in the F-phase, suggesting that the dynamical magnetic properties of metallic Mn_3Pt in the F-phase could not be fully understood under the molecular-field approximation with constant exchange interaction, and the itinerant nature of the magnetic electron should be taken into account. Further study on the temperature dependence of the spin wave will be needed by using the same sample to rule out the capability of the sample dependence.

Additional gapless mode found in the high energy-resolution measurement is consistent to the magnetic structure of the F-phase. The existence of the two acoustic modes, gapped and gapless mode, was proposed in the partial disordered state [15]. The gapless mode corresponds to the rotation of all the spins simultaneously by the same amount in the *bc*-plane in fig. 1(a). Another magnetic structure for the high-temperature phase was proposed by Long [16]. This is a non-collinear magnetic structure like the D-phase. All the Mn moments are ordered along twelve [110] directions. Neutron diffraction measurements cannot distinguish these two magnetic structures, because they have identical magntic peak positions and intensities. It was proposed that three acoustic modes will be exist in this non-collinear structure, and they are all gapped [15]. The existence of the gapless mode is characteristic of the collinear structure, and therefore, our result suggests that the magnetic structure is indeed the partial disordered state, in addition to the fact of the tetragonal distortion in the F-phase [6].

5. Summary

Spin wave dispersion in the F-phase just above the transition temperature was investigated by using the neutron scattering technique. The spin wave stiffness at $T = 400$ K is 50% harder than that at $T = 419$ K [10], while the spin wave energy gap was unchanged. This indicates the itinerancy of the magnetic moments in the F-phase. In addition, another gapless mode was found, indicating that the magnetic structure of the high-temperature phase is indeed the partial disordered state.

Acknowledgments

One of the author (S. Ibuka) is indebted to Y. Ishikawa and K. Namba at KEK for supporting the chemical analysis. Part of this work was performed using the facilities of CROSS-Tokai user laboratories, and the facilities of the Institute for Solid State Physics, the University of Tokyo. The neutron experiment at the Materials and Life Science Experimental Facility of J-PARC was performed under a user program (Proposal No. 2014B0153).

References

[1] Kamihara Y, Watanabe T, Hirano M and Hosono H 2008 *J. Am. Chem. Soc.* **130** 3296–3297
[2] Kohgi M and Ishikawa Y 1980 *J. Phys. Soc. Jpn.* **49** 985–993
[3] Ishikawa Y 1978 *J. Appl. Phys.* **49** 2125–2130
[4] Krén E, Kádár G, Pál L and Szabó P 1967 *J. Appl. Phys.* **38** 1265–1266
[5] Krén E, Kádár G, Pál L, Sólyom J, Szabó P and Tarnóczi T 1968 *Phys. Rev.* **171** 574–585
[6] Ikeda T and Tsunoda Y 2003 *J. Phys. Soc. Jpn.* **72** 2614–2621
[7] Yasui H, Ohashi M, Abe S, Yoshida H, Kaneko T, Yamaguchi Y and Suzuki T 1992 *J. Magn. Magn. Mater.* **104-107** 927–928
[8] Ricodeau J A 1974 *J. Phys. F: Metal Phys* **4** 1285–1303
[9] Tomiyoshi S, Yasui H, Kaneko T, Yamaguchi Y, Ikeda H, Todate Y and Tajima K 1990 *J. Magn. Magn. Mater.* **90 & 91** 203–204
[10] Yamaguchi Y, Yasui H, Funahashi S, Yamada M, Ohashi M and Kaneko T 1992 *Physica B* **180 & 181** 235–237
[11] Tomiyoshi S, Kaneko T, Steigenberger U, Chapell A J, Hagen M and Todatede Y 1992 *Physica B* **180 & 181** 227–229

[12] Ikeda T and Tsunoda Y 2004 *J. Magn. Magn. Mater.* **272-276** 482–484
[13] Tomiyasu K, Yasui H and Yamaguchi Y 2012 *J. Phys. Soc. Jpn.* **81** 114724–1–4
[14] Azuah R T, Kneller L R, Qiu Y, Tregenna-Piggott P L W, Brown C M, Copley J R D and Dimeo R M 2009 *J. Res. Natl. Inst. Stan. Technol.* **114** 341
[15] Hirano Y, Suzuki N, Shirai M and Motizuki K 1995 *J. Magn. Magn. Mater.* **140-144** 1975–1976
[16] Long M W 1991 *J. Phys.: Condens. Matter* **3** 7091–7094

Magnetic properties of weak itinerant electron ferromagnet CrAlGe

Soshi Yoshinaga[1], Yoshifuru Mitsui[1], Rie Y. Umetsu[2], Yoshiya Uwatoko[3], Keiichi Koyama[1]

[1] *Graduate School of Science and Engineering, Kagoshima University, Kagoshima 890-0065, Japan*
[2] *Institute for Materials Research, Tohoku University, Sendai 980-8577, Japan*
[3] *Institute for Solid State Physics, The University of Tokyo, Kashiwa 277-8581, Japan*

E-mail: k5997477@kadai.jp

Abstract. Magnetic properties for CrAlGe with an orthorhombic $TiSi_2$-type structure were investigated under ambient pressure and hydrostatic pressure. CrAlGe is ferromagnetic at the temperature below Curie temperature T_C of 80 K and has the spontaneous magnetic moment p_s of 0.41 μ_B/f.u. at 5 K and the effective moment p_{eff} of 1.89 μ_B/f.u. under ambient pressure. For the temperature T below 30 K, the decrease in $p_s(T)^2/p_s(5K)^2$ was proportional to T^2. However, the decrease in $p_s(T)^2/p_s(5K)^2$ was proportional to $T^{4/3}$ for 30 K < T < T_C. The ratio p_{eff}/p_s was consistent with the theoretical curve of the generalized Rhodes-Wohlfarth plot (Takahashi plot). Obtained results suggest that CrAlGe is a weak itinerant electron ferromagnet. By applying a high pressure of 1 GPa, T_C decreased by approximately 10 K.

1. Introduction

It was reported that ferromagnetic MnAlGe with the tetragonal Cu_2Sb-type structure is one of the promising candidates for the application of high-density memory and spintronics devices [1]. In this crystal structure, as shown in Fig. 1(a), Mn atoms occupy the 2a-site that locates two-dimensionally at c-plane, resulting a large uniaxial magnetic anisotropy of about 10^6 J/m^3 along c-axis [2-4]. The spontaneous magnetic moment p_s is 1.70 μ_B/f.u. and the Curie temperature T_C is 518 K [2-4]. It was reported that Cr substitution for Mn in $Mn_{1-x}Cr_xAlGe$ (x $\leq$ 0.2) with the Cu_2Sb-type structure enhanced p_s and T_C [3, 4]. However, it has been unclear why the magnetic properties improve by Cr substitution for Mn in $Mn_{1-x}Cr_xAlGe$. In order to understand the Cr substitution effect on the magnetic properties of $Mn_{1-x}Cr_xAlGe$, it is important to study the fundamental properties of CrAlGe as well as $Mn_{1-x}Cr_xAlGe$ systematically. However, the structural and magnetic properties of CrAlGe had been not characterized in detail, whereas ternary phase diagram of Cr-Al-Ge at 673 K shows existence of CrAlGe phase as an equilibrium state [5, 6]. Quite recently, Yoshinaga *et al.* suggested that CrAlGe has not the Cu_2Sb-type but the orthorhombic $TiSi_2$-type structure, as shown in Fig. 1(b). Furthermore, they reported that CrAlGe with the $TiSi_2$-type structure is a weak itinerant electron magnet [7].

In this paper, the experimental results on the magnetic properties of polycrystalline CrAlGe is presented. The magnetic data suggest that the Curie temperature and the magnetic moment of CrAlGe are suppressed by applying high pressure. The results obtained under ambient pressure are discussed based on Takahashi's spin fluctuation theory.

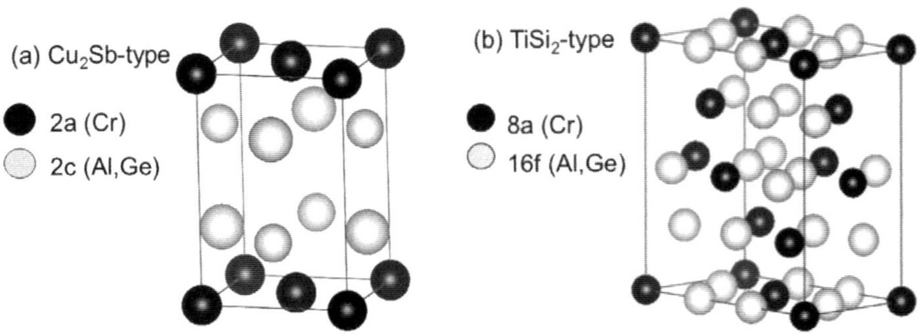

Figure 1. Crystal structures of tetragonal Cu_2Sb-type (a) and orthorhombic $TiSi_2$-type (b).

2. Experimental procedure

Polycrystalline CrAlGe was prepared by induction melting of a mixture composed from appropriate amounts of pure elements (Cr, 99.99%; Al, 99.99%; Ge, 99.99%). The as-cast sample was annealed at 1073 K for 1 day and then quenched into iced water. Powder x-ray diffraction (XRD) experiments were carried out using Cu-Kα radiation at room temperature. For determining the crystal structure and the x-ray reflection indices, the observed diffraction patterns were analyzed by comparison with calculated patterns using a Powder Cell program [8]. The magnetization M measurements under ambient pressure were carried out using a superconducting quantum interference device (SQUID) magnetometer (Quantum Design) for $5 \leq T \leq 390$ K and magnetic fields H up to 5 T at IMR. The magnetization under a hydrostatic pressure of 1 GPa was measured using the SQUID magnetometer and a clamp-type piston cylinder pressure cell at ISSP.

3. Results and discussion

Figure 2 shows the experimental XRD pattern (a) and the calculated XRD pattern for the orthorhombic $TiSi_2$-type structure (b) [7]. The parameters used for the calculated XRD pattern of the $TiSi_2$-type structure are listed in Table 1. In this study, we assumed that Cr atom occupied the 8a-site, and both atoms of Al and Ge were at the 16f-site by the same occupancy in the $TiSi_2$-type structure. As seen in this Fig. 2, the calculated XRD is in good agreement with the experimental XRD pattern (a). All the diffraction peaks were indexed as a single phase of the $TiSi_2$-type structure. The lattice parameters were determined to be $a = 0.4770$ nm, $b = 0.8254$ nm and $c = 0.8725$ nm.

Figure 3 shows the magnetization curves in magnetic fields up to 5 T (a) and the isothermal M^2 *vs.*

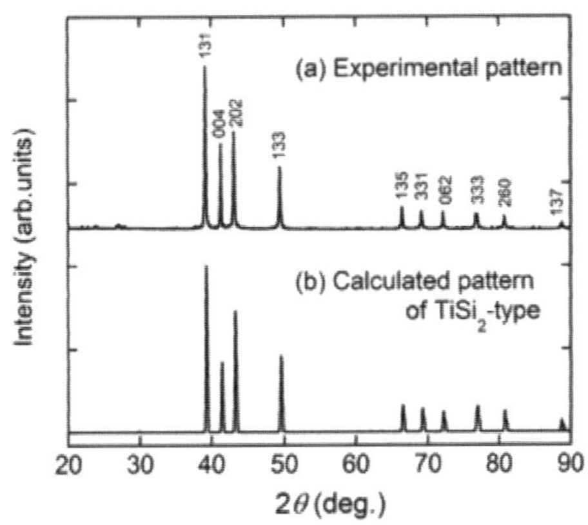

Figure 2. Experimental XRD pattern (a) and the calculated XRD pattern for the orthorhombic $TiSi_2$-type structure (b) [7].

IOP Conf. Series: Journal of Physics: Conf. Series **868** (2017) 012018 doi:10.1088/1742-6596/868/1/012018

Table 1. Used parameters of $TiSi_2$-type structures for CrAlGe [7].

Atom	*Wyckoff*	*x* (nm)	*y* (nm)	*z* (nm)	*Occupancy*
Cr	8a	0	0	0	1
Al	16f	0	0.3365	0	0.5
Ge	16f	0	0.3365	0	0.5

H/*M* plots (Arrott plot) (b) the at several temperatures. The magnetization curves for $T \leq 80$ K show the ferromagnetic behaviour. The spontaneous magnetic moment p_s at 5 K was determined to be 0.41 μ_B/f.u. (0.41 μ_B/Cr) by Arrott plot. We estimated p_s^2 at several temperatures from the Arrott plot analysis for $T \leq 90$ K, as shown in Fig. 3 (b). The Curie temperature T_C was determined to be 80 K as the temperature at which p_s^2 passes the origin, as shown in Fig. 4.

Figure 5 shows the temperature dependence of the reciprocal susceptibility $1/\chi$ ($1/\chi$ *vs*. *T* plot) for $T > 150$ K [7]. The susceptibility χ was deduced from the temperature dependence of the magnetic moment in a magnetic field of 5 T. The observed $1/\chi$ *vs*. *T* plot is well expressed by the Curie-Weiss

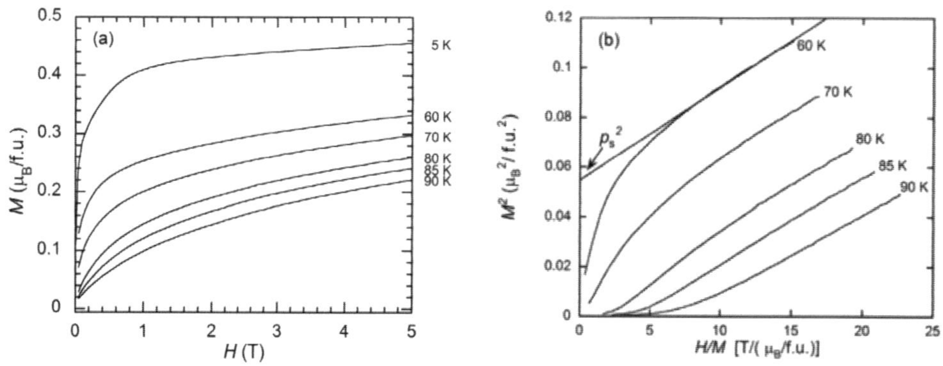

Figure 3. Magnetization curves (a) and M^2 *vs*. *H*/*M* plots (Arrott plot) (b) of CrAlGe at several temperatures. Spontaneous magnetic moment p_s was determined by the linear extrapolation to $H/M = 0$ for the M^2 *vs*. *H*/*M* plots.

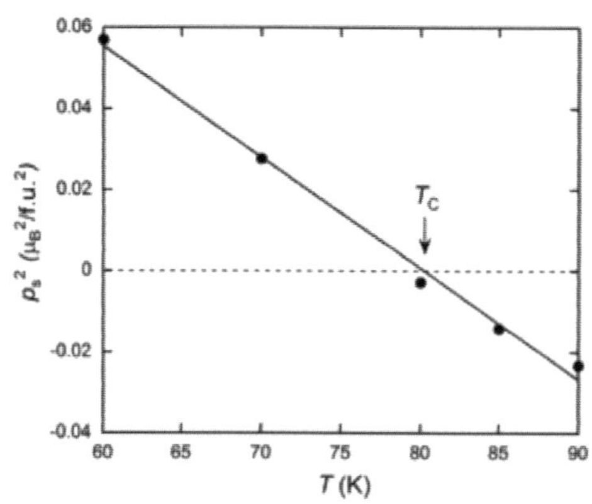

Figure 4. p_s^2 *vs*. *T* plot for $60 \leq T \leq 90$ K. The solid line is least-squares fits to p_s^2.

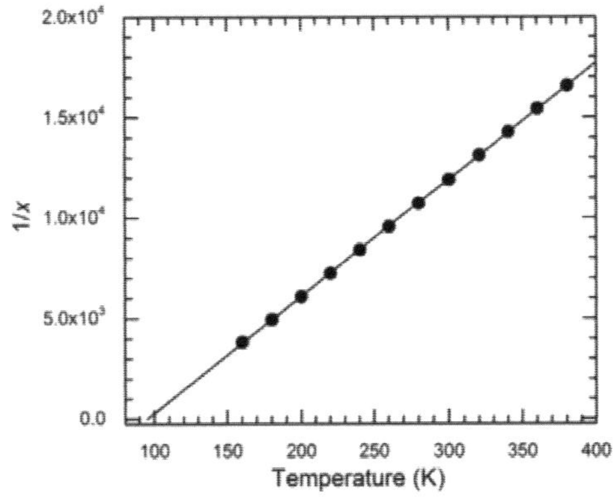

Figure 5. Temperature dependence of the reciprocal susceptibility $1/\chi$ [7].

law, and the paramagnetic Curie temperature θ_p and the effective magnetic moment p_{eff} were determined to be 95 K and 1.89 μ_B/Cr, respectively.

The Curie-Weiss law is concluded from the Weiss theory (called localized electron model) which assumes a fixed magnetic moment per atom [9]. The effective magnetic moment p_{eff} deduced from the Curie constant and the spontaneous magnetic moment per atom of the ground state p_s have the following relation:

$$p_{\mathrm{eff}}^2 = g^2 S(S+1), \qquad p_s = gS, \tag{1}$$

where g and S are g-factor and the spin angular momentum, respectively. It is well known that the experimental results on many magnetic metals and alloys, which are not represented based on the localized electron model, also show the Curie-Weiss low. In addition, the value of p_c is calculated from p_{eff} using $p_{eff}^2 = p_c(p_c+2)$, and in general, p_c/p_s for ferromagnetic metals with high Curie temperature is nearly unity [9]. On the other hand, p_c/p_s for weak itinerant-electron-ferromagnets (WIEF) with low Curie temperature becomes very large [9].

In this study, it was found that CrAlGe had magnetic properties with T_C = 80 K, θ_p = 95 K, p_s = 0.41 μ_B/Cr at 5 K and p_{eff} = 1.89 μ_B/Cr. That is, p_c/p_s is larger than unity, indicating that CrAlGe is one of WIEF. In order to clarify the origin of the magnetic properties of CrAlGe, it is important to analyze the magnetic data based on Takahashi's spin fluctuation theory for WIEF [10-12]. Therefore, we estimated the paramcters T_0 and T_A which arc characterized by the energy width of the dynamical spin fluctuation spectrum and the dispersion of the static magnetic susceptibility in the wave vector space, respectively. The quantitative values of T_0 and T_A can be deduced from the experimental data of the Arrott plot on the basis of Takahashi's theory. The analysis method for CrAlGe was presented in the previous paper in detail [7]. These parameters were estimated to be $T_0 = 1.0 \times 10^3$ K, $T_A = 4.0 \times 10^3$ K and η (= $(T_C/T_0)^{1/3}$) = 0.43 by the magnetization process (Arrott plot) at 5 K. These values are comparable to those of typical WIEF such as MnSi, $ZrZn_2$, Ni_3Al, (Fe, Co)Si, etc. [12].

Table 2. The parameters estimated from magnetization measurements.

	p_s (μ_B)	p_{eff} (μ_B)	p_c (μ_B)	p_{eff}/p_s	p_c/p_s	T_C (K)	T_0 (K)	T_C/T_0
CrAlGe	0.413	1.89	1.13	4.58	2.75	80	1028	0.078

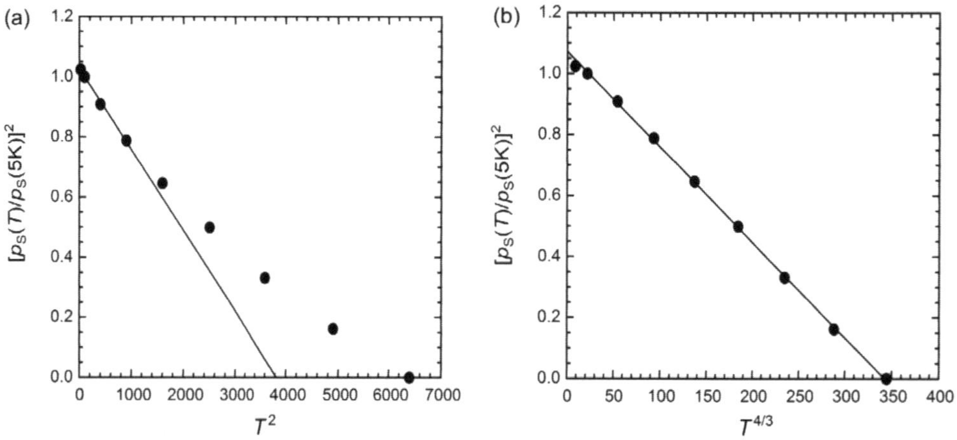

Figure 6. $[p_s(T)/p_s(5K)]^2$ *vs.* T^2 plot (a) and $[p_s(T)/p_s(5K)]^2$ *vs.* $T^{4/3}$ plot (b) of CrAlGe [7]. The straight lines are least-squares fits to the data.

The obtained η (= 0.43) of CrAlGe was small, which is consistent with Takahashi's theory for WIFE [12]. In this case, according to Takahashi's theory, $[p_s(T)/p_s(0)]^2$ follows the $T^{4/3}$–dependence in the temperature range below T_C except for very low temperature at which the T^2–dependence is observed. In Fig. 6(a) and (b), $[p_s(T)/p_s(5K)]^2$ *vs.* T^2 and $T^{4/3}$ plots are shown, respectively. As seen in Fig. 6(a) and (b), $[p_s(T)/p_s(0)]^2$ shows a T^2–dependence at low temperature side and a $T^{4/3}$–dependence in the wide temperature range below T_C. By using the low temperature data for $5 \leq T \leq 30$ K, T_A is estimated to be 3.8×10^3 K, which is in good agreement with the value (4.0×10^3 K) determined by the magnetization process at 5 K, as mentioned above. Consequently, the magnetization process of CrAlGe at 5 K and the temperature dependence of the spontaneous magnetic moment are consistent with Takahashi's theory. The obtained values of the parameters are summarized in Table 2.

It is reported that the small amount of Cr substitution for Mn atom of MnAlGe with the tetragonal Cu_2Sb-type structure enhanced p_s and T_C [3, 4]. However, p_s (= 0.413 μ_B/f.u.) and T_C (= 80 K) of fully substituted compound CrAlGe with the orthorhombic $TiSi_2$-type were much lower than those (p_s = 1.70 μ_B/f.u. and T_C = 518 K) of MnAlGe with the tetragonal. This is probably due to the change of the crystal structure. That is, it is supposed that the unique properties of $Mn_{1-x}Cr_xAlGe$ originate from the magnetic characters of Mn and Cr atoms located two-dimensionally in c-plane of the tetragonal Cu_2Sb-type structure.

It is well known that Rhodes-Wohlfarth plot (p_c/p_s *vs.* T_C plot) is able to categorize the ferromagnets which are the localized electron model and the itinerant electron model [13]. In addition to this, recently, p_{eff}/p_s *vs.* T_C/T_0 plot (Takahashi plot) was also proposed by Takahashi.

Figure 7 shows the Rhodes-Wohlfarth plot (a) and the Takahashi plot (b), respectively. The solid curve in Fig.8 (b) is the theoretical curve. As seen in Fig. 8 (a) and (b), CrAlGe is not a localized electron ferromagent but WIEF, and the position of CrAlGe is in agreement with the curve in the Takahashi plot (b), compared with the curve in the Rhodes-Wohlfarth plot (a). This result suggests that the magnetic properties of CrAlGe depend strongly on the parameters proposed in Takahashi's spin fluctuation theory. In other words, the magnetic properties of CrAlGe originate the quantum spin fluctuation. As seen in Fig. 8 (b), however, it seems that the position deduced from our experimental results for CrAlGe is slightly deviated from the theoretical curve in Takahashi plot. This is may be due to the magnetic anisotropy of the sample. In the case of WIEF with the magnetic anisotropy, the parameters T_0 and T_A have to be modified slightly.

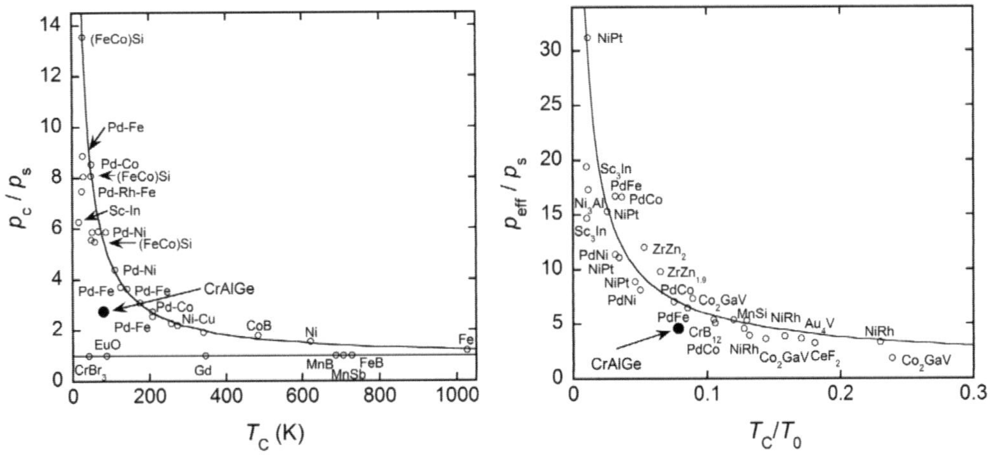

Figure 7. Rhodes-Wohlfarth plot (a) and generalized Rhodes-Wohlfarth plot (Takahashi plot) (b). Closed circle and open circles indicate the data of this work for CrAlGe and reported experimental data [12], respectively.

Finally, we present the pressure effect on the magnetic properties of CrAlGe. Figure 8 shows the temperature dependence of the magnetization (*M-T* curve) at a magnetic field of 1 T under 0.1 MPa (ambient pressure) and 1 GPa. In these measurements, T_C was defined as the peak of the temperature derivative of *M-T* curves (dM/dT). As seen in Fig. 8, *M* at 5 K and T_C decreased by about 5% and 10 K, respectively, when a high pressure of 1 GPa was applied. As discussed above, the magnetic properties of CrAlGe originate from the nature of WIEF. When a hydrostatic pressure was applied to CrAlGe, the cell volume decreases. If the electronic band structure of CrAlGe represents the magnetic properties at 0 K (lowest temperature), this volume contraction leads the increase of the bandwidth and the density of states in the vicinity of the Fermi level. This change of the band structure probably causes the enhancement of the spin fluctuation and the decrease of *M* and T_C.

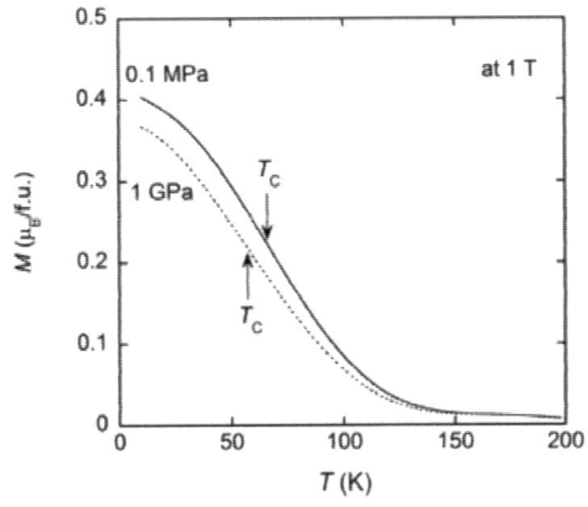

Figure 8. Temperature dependence of the magnetization at a magnetic field of 1 T under 0.1 MPa and 1 GPa.

4. Conclusions

CrAlGe has an orthorhombic $TiSi_2$-type structure with the lattice parameters of a = 0.4770 nm, b = 0.8254 nm and c = 0.8725 nm at room temperature. The spontaneous magnetic moment, effective

IOP Conf. Series: Journal of Physics: Conf. Series **868** (2017) 012018 doi:10.1088/1742-6596/868/1/012018

magnetic moment and Curie temperature of CrAlGe were determined to be $p_s = 0.41$ μ_B/Cr at 5 K, $p_{eff} = 1.89$ μ_B/Cr and $T_C = 80$ K, respectively. By analyzing magnetic data, the parameters T_0 and T_A were estimated to be 1.0×10^3K and 4.0×10^3K, respectively. The ratio p_{eff}/p_s was consistent with the theoretical curve of the generalized Rhodes-Wohlfarth plot (Takahashi plot: p_{eff}/p_s *vs.* T_C/T_0 plot). Obtained results suggested that CrAlGe is a weak itinerant electron ferromagnet expressed well by Takahashi's spin fluctuation theory. It was found that the magnetic moment and the Curie temperature decreased by applying high pressure.

Acknowledgements
Magnetization measurements under ambient pressure were performed at Low-temperature center, Institute for Materials Research, Tohoku University. The experiments under high pressures were carried out at Institute for Solid State Physics, the University of Tokyo.

References
[1] S. Mizukami, A. Sakuma, T. Kubota, Y. Kondo, A. Sugihara and T. Miyazaki, Appl. Phys. Lett.103, 142405 (2013).
[2] J. H. Wernick, S. E. Haszko and W. J. Romanov, J. Appl. Phys. 32, 2495 (1961).
[3] H. Ido, T. Kamimura and K. Shirakawa, J. Appl. Phs. 55, 2365 (1984).
[4] R. Y. Umetsu, Y. Mitsui, I. Yuito, T. Takeuchi and H. Kawarada, IEEE Trans. Magn. 11, 1001904 (2014).
[5] AtomWork <http://crystdb.nims.go.jp/>
[6] Y. Xu, M. Yamazaki and P. Villars, Jpn. J. Appl. Phys. 50, 11RH02 (2011).
[7] S. Yoshinaga, Y. Mtisui, R. Y. Umetsu and K. Koyama, Physics Procedia, in press.
[8] W. Kraus and G. Nolze, J. Appl. Cryst, 29 301 (1996).
[9] S. Chikazumi, *Physics of Ferromagnetism* (International Series of Monographs on Physics vol.94, Oxford Science Publications, Oxford, 2009).
[10] Y. Takahashi, J. Phys. Soc. Jpn. 55, 3553 (1986).
[11] Y. Takahashi, J. Phys.: Condens. Matter 13, 6323 (2001).
[12] Y. Takahashi, *Spin Fluctuation Theory of Itinerant Electron Magnetism* (Springer Tracts in Modern Physics vol.253, Springer Berlin Heidelberg, 2013)
[13] P. Rhodes and E. P. Wohlfarth, Proc. R. Soc. 273, 247 (1963)

IOP Conf. Series: Journal of Physics: Conf. Series **868** (2017) 012019 doi:10.1088/1742-6596/868/1/012019

Drastic effect of the Mn-substitution in the strongly correlated semiconductor $FeSb_2$.

Mohamed A. Kassem*, Yoshikazu Tabata, Takeshi Waki, Hiroyuki Nakamura.

Department of Materials Science and Engineering, Kyoto University, Kyoto 606-8501, Japan.

E-mail: kassem.ahmed.82s@st.kyoto-u.ac.jp

Abstract. We report the effects of Mn substitution, corresponding to hole doping, on the electronic properties of the narrow gap semiconductor, $FeSb_2$, using single crystals of $Fe_{1-x}Mn_xSb_2$ grown by the Sb flux method. The orthorhombic *Pnnm* structure was confirmed by powder X-ray diffraction (XRD) for the pure and Mn-substituted samples. Their crystal structure parameters were refined using the Rietveld method. The chemical composition was investigated by wavelength-dispersive X-ray spectroscopy (WDX). The solubility limit of Mn in $FeSb_2$ is x_{max} ~ 0.05 and the lattice constants change monotonically with increasing the actual Mn concentration. A drastic change from semiconducting to metallic electronic transports was found at very low Mn concentration at x ~ 0.01. Our experimental results and analysis indicate that the substitution of a small amount of Mn changes drastically the electronic state in $FeSb_2$ as well as the Co-substitution does: closing of the narrow gap and emergence of the density of states (DOS) at the Fermi level.

1. Introduction

Strongly correlated electron materials have unusual electronic structures and show interesting properties, such as metal-insulator transitions, heavy fermions, narrow-gap semiconductors and high-T_c superconductivity. Among of them, the narrow-gap semiconductors, such as Kondo semiconductors are characterized by a narrow gap at the Fermi level and sharp DOS peaks near the Fermi level due to the electron correlations and hybridization between broad conduction electron bands and narrow *d*- or *f*-electron bands [1].

FeSi and $FeSb_2$ are a few candidates of *d*-electron based electron-correlation driven semiconductors [1,2]. In recent years considerable attention has been paid for studying $FeSb_2$ because of its anomalously huge Seebeck coefficient [3]. $FeSb_2$ crystallizes in the marcasite crystal structure of orthorhombic space group *Pnnm* with lattice parameters of a = 5.8328(5), b = 6.5376(5) and c = 3.1973(3) Å [4]. Fe-atom at (0, 0, 0) is octahedrally surrounded by Sb atoms at (x, y, 0) and the $FeSb_6$ octahedra forms edge-shared chains along the c axis. Two-energy-gap behavior in resistivity and thermally activated Pauli paramagnetic behavior were experimentally observed in single crystals of $FeSb_2$ [5,6]. Evolution of ferromagnetic ground state and gap closing were reported by the substitution of a small amount of third element, for instance, by substituting Co for Fe and Te for Sb [7,8]. The high sensitivity on its transport and magnetic properties against element substitutions is important to understand the nature of the narrow-gap semiconducting state of $FeSb_2$ because it can be explained based on the electron-correlation driven mechanism.

* Permanent Address: Department of Physics, Assiut University, Assiut 71515, Egypt

In this paper we report on the Mn-substitution effects on the electronic transport properties of $FeSb_2$, which is a counterpart of the Co-substitution effects: the former is a hole doping and the latter is an electron doping. The systematic substitution of Mn for Fe and its solubility limit were investigated by XRD and SEM-WDX for pure and doped samples. The electrical resistivity measurements show drastic changes in the electronic state and gap closing in $Fe_{1-x}Mn_xSb_2$. Interestingly, the energy gap is much more sensitive to the Mn-substitution than the Co-substitution.

2. Experimental Details

Silver rod-shaped single crystals of $Fe_{1-x}Mn_xSb_2$ with the long axis parallel to the b axis were synthesized by the high temperature self Sb-flux method. The grown crystals were characterized by powder XRD at room temperature (PANalytical, X'Pert Pro) with Cu$K_{\alpha 1}$ radiation monochromated by a Ge (111)-Johansson-type monochromator. The crystal structure parameters were refined by the Rietveld analysis using TOPAS (Bruker) [9]. To identify the actual Mn concentrations in the grown crystals, the chemical compositions were investigated by SEM-WDX (Hitachi, S-3500H). The actual Mn-concentrations measured by the WDX were found to be considerably reduced from the nominal starting concentrations. The WDX results also indicated the solubility limit of $x_{\max} \approx 0.05$. The crystalline axes were determined by the X-ray Laue method. The susceptibility of the parent compound, $FeSb_2$, was measured using a SQUID magnetometer (Quantum Design, MPMS-7) in a temperature range of $T = 2$ - 350 K by applying magnetic field along the three axes. Unfortunately, a very small amount of ferromagnetic impurity with $T_C \sim 150$ K, which is too small to be detected by any other techniques, masks the intrinsic susceptibility in the Mn-substituted $Fe_{1-x}Mn_xSb_2$. Hence, we do not discuss the magnetic properties except for pure $FeSb_2$. Electrical resistivity was measured by the four probe method of rectangular rods of single crystals cut along the three principal crystal axes in a temperature range of $T = 4$ - 300 K.

3. Results and Discussions

The phase and the crystal structure are investigated by powder XRD using powders crushed from the crystals. The observed diffraction patterns are fitted by the Rietveld method using the structure parameters of $FeSb_2$ marcasite structure of *Pnnm* orthorhombic symmetry. Both observed and calculated XRD patterns of $Fe_{1-x}Mn_xSb_2$ at room temperature as well as the differences are shown in Fig. 1. The main phase of the obtained crystals for all Mn concentrations is $FeSb_2$ phase. Figure 2 shows the refined lattice parameters a, b and c (Fig. 2a) and the unit cell volume V (Fig. 2b) plotted against Mn concentration x. The lattice parameters as well as unit cell volume increase linearly as x increasing and obey well the Vegard's law indicating successful substitution of Mn for Fe in $Fe_{1-x}Mn_xSb_2$. The crystal structure parameters with the reliability R-factors and goodness-fit-indicator S for different Mn concentrations are listed in Table 1. The Fe/Mn occupancies were fixed based on the WDX results. The low values of S and R-factors indicate the satisfactory refinement.

Figure 3 shows the temperature dependences of the magnetic susceptibility M/H of $FeSb_2$ single crystals with H parallel to the three principal crystal axes, where M is the magnetization and H is the applied magnetic field. A thermally activated behavior was found as well as previous reports [2,6]. Upturns at low T are attributed to small amounts of magnetic impurities. Following the previous reports [6,7], we analyze the temperature dependence of the magnetic susceptibility based on the narrow-band small-gap model with a density of states $N(E)$ consisting of two narrow bands of the width W separated by a gap $E_g = 2\Delta$. In this model, described in the inset of Fig. 3, the functional form of the magnetic susceptibility is written as, $\chi(T) = \chi_0 - 2\mu_B^2 \int N(E) \frac{\partial f(E,T)}{\partial E} dE$. Least square fitting of the polycrystalline average, $\chi_{avg.} = (\chi_a + \chi_b + \chi_c)/3$ yields $\Delta = 35.76 \pm 0.41$ meV, $W = 27.91 \pm 1.35$ meV and $\chi_0 = 2.2 \times 10^{-5}$ emu/mole which agree well with the reported results [6,7]. Figure 4 shows the electrical resistivity $\rho(T)$ of $FeSb_2$ single crystals along the three principal crystal axes against temperature. Qualitatively identical semiconducting behaviors were observed along the three axes up to 300 K.

IOP Conf. Series: Journal of Physics: Conf. Series **868** (2017) 012019 doi:10.1088/1742-6596/868/1/012019

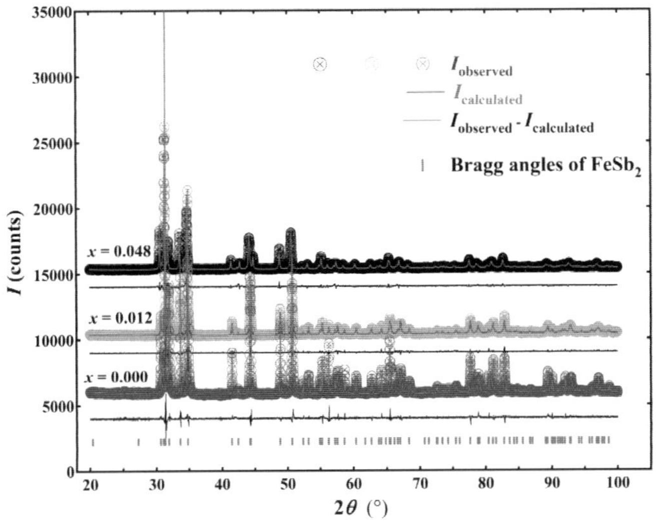

Figure 1. (Color online) Powder XRD patterns of $Fe_{1-x}Mn_xSb_2$ for representative x with the fitting results using the Rierveld method. The differences between experimental and fitting results are denoted in black solid lines. The positions of Bragg angles for $FeSb_2$ are indicated by red vertical bars. For clear visibility, the XRD patterns of x = 0.012 and x = 0.048 are vertically shifted.

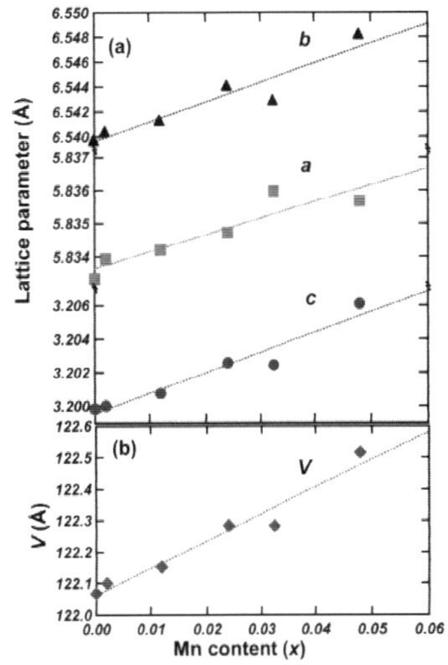

Figure 2. Mn concentration dependence of (a) the lattice parameters and (b) unit cell volume of $Fe_{1-x}Mn_xSb_2$.

Table 1: Refinement parameters, lattice parameters and atomic position parameters of Sb, x and y, refined by the Rietveld analysis of XRD patterns at room temperature using TOPAS. The marcasite phase with space group *Pnnm* and atomic positions of Fe/Mn at (0,0,0) and Sb at (x,y,0) are assumed (see text).

	x = 0.0	x = 0.002	x = 0.012	x = 0.024	x = 0.032	x = 0.048
R_{exp}	2.88	3.04	4.29	3.87	3.16	4.41
R_{wp}	4.44	7.66	7.65	9.02	7.53	8.76
R_p	3.40	5.33	5.87	6.30	5.33	6.46
$S= R_{wb}/R_{exp}$	1.54	2.52	1.78	2.33	2.38	1.99
a	5.8333(3)	5.834(3)	5.834(3)	5.835(4)	5.836(1)	5.836
b	6.5396(4)	6.540(3)	6.541(4)	6.543(5)	6.543(1)	6.548
c	3.1998(2)	3.200(2)	3.201(1)	3.202(2)	3.202(1)	3.206
x	0.1882(1)	0.1887(3)	0.1897(3)	0.1918(4)	0.1903(5)	0.18977
y	0.35582(9)	0.3565(2)	0.3563(2)	0.3564(3)	0.3555(3)	0.35608

The observed saturating behavior below 5 K and the anomaly at ~ 40 K are probably extrinsic due to a very small amount of impurities. As reported previously [3,5], two energy-gap behavior was observed. This two-gap behavior is interpreted as a manifestation of a temperature-dependent DOS, a gap-forming, in the electron-correlation driven semiconductor. The energy gaps ε_{g1} and ε_{g2} are estimated by the linear fits in the ln(ρ) vs $1/T$ plot (Arrhenius plot), shown in the inset of Fig. 4, in the temperature

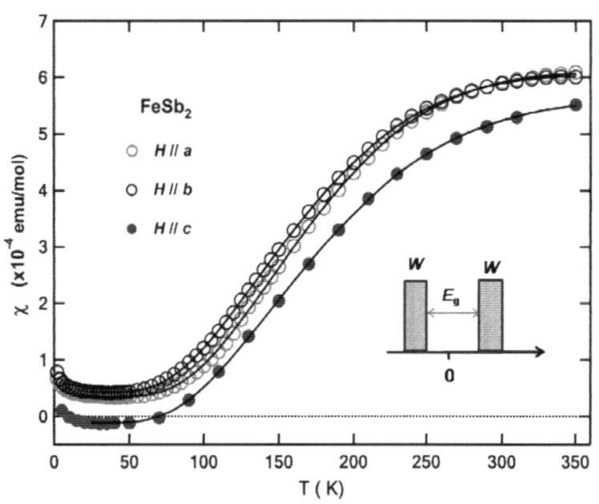

Figure 3. Temperature dependences of the magnetic susceptibility of $FeSb_2$ measured at $H = 0.1$ T applied along principal axes. The solid lines are fitting results based on the narrow-band small-gap model. Inset shows a schematic drawing of the DOS of the model.

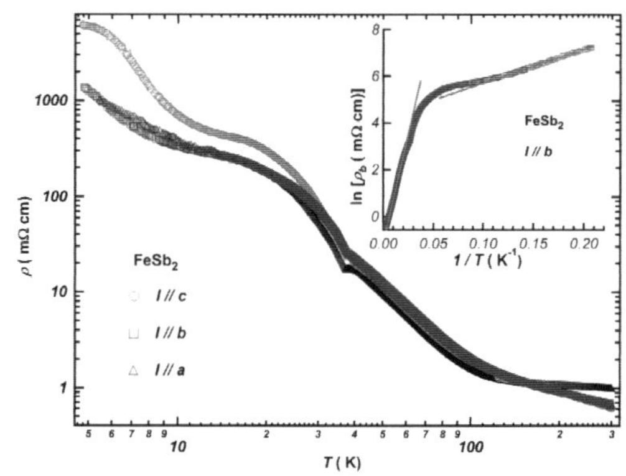

Figure 4. Electrical resistivity $\rho(T)$ of $FeSb_2$ for current applied along the three crystal axes. Inset shows the Arrhenius plot of $\rho_b(T)$ data in b direction and solid lines represent the linear fits (see text).

ranges $T = 5$ - 10 K and $T = 50$ - 120 K, respectively. The obtained values of ε_{g1} and ε_{g2} are about 2.5 meV and 33 meV, respectively, which are consistent with reported values [3,5]. The observed magnetic and electronic transport properties indicate a quality of our single crystals of $FeSb_2$ as high as in refs. 3, 5, 6.

We show the temperature dependences of the electrical resistivities of $Fe_{1-x}Mn_xSb_2$ single crystals along the b-axis in the form of Arrhenius plot in Fig. 5. The resistivity ρ_b is strongly reduced with increasing of the Mn-concentration at low temperatures. Especially, drastic change of ρ_b is found around $x \sim 0.01$, where ρ_b is reduced in two orders of magnitude. We estimate the gap energies ε_{g1} and ε_{g2} at low and high temperatures, respectively, by linear fitting. The slope in the high temperature region is almost not affected by the Mn-substitution, whereas the slope at low temperature is reduced continuously and becomes negative with increasing the Mn concentration. These variations indicate

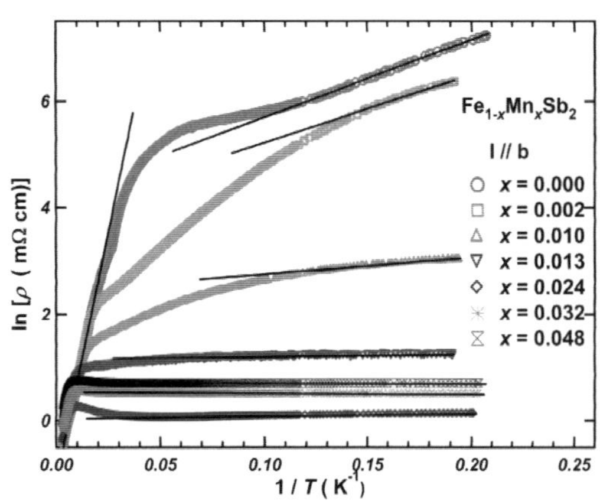

Figure 5. Electrical resistivity presented as Arrhenius plots of $Fe_{1-x}Mn_xSb_2$ single crystals for current applied along b direction.

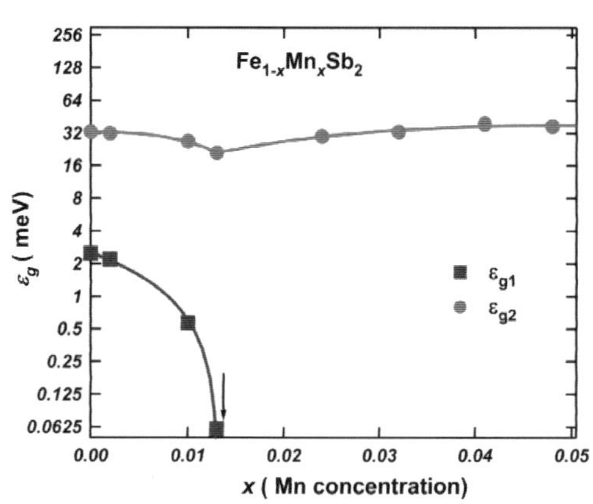

Figure 6. Mn concentration-dependence of the two energy gaps ε_{g1} and ε_{g2} in $Fe_{1-x}Mn_xSb_2$. Solid lines are guides to eyes and the vertical arrow indicates the critical Mn concentration of the small gap closing.

the drastic change of the resistivity is attributed to continuous reduction and disappearance of the smaller energy gap ε_{g1}. The negative slope above $x > 0.013$ indicate an emergence of the density of states (DOS) at the Fermi level and a metallic ground state upon the Mn-substitution. Estimated ε_{g1} and ε_{g2} are plotted against the Mn-concentration in Fig. 6. The small energy gap ε_{g1} decreases upon the Mn-substitution and closes completely just above $x = 0.013$ before the emergence of the metallic state at higher Mn concentrations. ε_{g2} slightly decreases first up to $x = 0.013$ then slightly increases by the Mn doping for samples showing metallic behavior at low temperatures. The observed electronic transport properties of $Fe_{1-x}Mn_xSb_2$ indicate that the metallic ground state induced by the Mn-substitution, corresponding to hole doping, in $FeSb_2$ semiconductor as well as by the electron-doping in the Co-substituted system, is attributed to the closing of the smaller energy gap ε_{g1}.

4. Conclusion

Single crystals of $FeSb_2$ and its Mn-substituted $Fe_{1-x}Mn_xSb_2$ for $x < 0.05$ were grown out of Sb flux. The successful and systematic Mn-substitution was indicated by XRD measurements and Rietveld analysis. Two energy-gap behavior in the resistivity and drastic suppression of the smaller gap by the Mn-substitution were found. The emergence of metallic state in $Fe_{1-x}Mn_xSb_2$ is due to the closing of the small energy gap of $FeSb_2$.

Acknowledgments

M.A.K. would like to thank the Ministry of Higher Education, Egypt for financial support during his PhD study.

References

[1] Aeppli G and Fisk Z 1992 *Comments Cond. Mat. Phys.* **16** 155
[2] Petrovic C, Kim J W, Bud'ko S L, Goldman A I, Canfield P C, Choe W and Miller G J 2003 *Phys. Rev.* B **67** 155205
[3] Bentien A, Johnsen S, Madsen G K H, Iversen B B and Steglich F 2007 *Europhys. Lett.* **80** 17008
[4] Holseth H and Kjekshus A 1969 *Acta Chem. Scand.* **23** 3043
[5] Sun P, Oeschler N, Johnsen S, Iversen B B and Steglich F 2010 *Dalton Trans.* **39** 1012
[6] Koyama T, Fukui Y, Muro Y, Nagao T, Nakamura H and Kohara T 2007 *Phys. Rev.* B **76** 073203
[7] Hu R, Mitrović V F and Petrovic C 2006 *Phys. Rev.* B **74** 195130
[8] Hu R, Mitrović V F and Petrovic C 2009 *Phys. Rev.* B **79** 064510
[9] Bruker AXS 2008 *TOPAS V4: General profile and structure analysis software for powder diffraction data. - User's Manual* (Karlsruhe: Bruker AXS)

IOP Conf. Series: Journal of Physics: Conf. Series **868** (2017) 012020 doi:10.1088/1742-6596/868/1/012020

Anomalous antiferromagnetic state in $Nd_2Co_{12}P_7$

Yusuke Watanabe[1], Hiroto Ohta[1], Hiroko Aruga Katori[1], Atsushi Miyake[2] and Masashi Tokunaga[2]

[1] Department of Applied Physics, Graduate School of Engineering, Tokyo University of Agriculture and Technology, Tokyo 184-8588, Japan
[2] Institute for Solid State Physics, The University of Tokyo, Chiba 277-8581, Japan

E-mail: `h-katori@cc.tuat.ac.jp`

Abstract. We synthesized polycrystalline and single crystal samples of $Nd_2Co_{12}P_7$ and measured magnetization of the samples to study the antiferromagnetic ground state of this compound. We con rmed that $Nd_2Co_{12}P_7$ changes from the ferromagnetic state to the antiferromagnetic one with the decrease in temperature without magnetic phase transitions. In the antiferromagnetic state, the value of the Nd sublattice moment is 1.25 μ_B per Nd at $H = 0$, and it reaches the value of 2 μ_B in the saturating state achieved by a high magnetic eld through a metamagnetic transition around 15 T. These values are smaller than the value expected for isolated Nd^{3+} ions, although the magnetic moment of Nd ions behaves like that of isolated Nd^{3+} ions in the paramagnetic state. The reason why the antiferromagnetic state realizes in the ferromagnetic state without magnetic phase transitions in $Nd_2Co_{12}P_7$ is that the internal magnetic eld derived from the ferromagnetic Co sublattice gradually induces the Nd sublattice moment with decreasing temperature and the values of both sublattice moments become the same around the ground state.

1. Introduction

Localized magnetic moments in itinerant electronic ferromagnets can induce anomalous magnetic phenomena. The most prominent examples are base materials of permanent magnets LnCo$_5$, Ln_2Co$_7$ and Ln_2Fe$_{14}$B (Ln = lanthanoids) etc. The spontaneous magnetization of the sublattice of the ferromagnetic transition metal in these materials is enhanced by localized magnetic moments of Ln in cases of light rare earth elements. Other examples are layered compounds LnCoPnO (Pn = P, As), which show ferromagnetic ordering of 3d electrons of Co below the Curie temperature T_C 70 K. Among LnCoPnO, anomalous antiferromagnetic states emerge in a low-temperature region of the ferromagnetic state in the cases of Ln = Nd, Sm, and Gd [1]. Although such antiferromagnetic transitions seem to be induced by localized magnetic moments of Ln [1], there have remained unsolved problems, e.g. the anomalous phase boundary between antiferromagnetic and ferromagnetic phases in the temperature-magnetic eld plane [2] and an anomalous temperature dependence of the Ln sublattice moments [3]. For understanding such unusual magnetic phenomena, we need to study various itinerant electronic ferromagnets with localized magnetic moments.

For this purpose, we focused on itinerant ferromagnets Ln_2Co$_{12}$P$_7$, which have $Zr_2Fe_{12}P_7$-type crystal structure (space group : $P\bar{6}$) as shown in Fig. 1 [4]. In these compounds, 3d electrons of Co show ferromagnetic ordering below T_C 150 K. In the ferromagnetic state, the evolution of the spontaneous magnetizations with the drop of the temperature depends on Ln

IOP Conf. Series: Journal of Physics: Conf. Series **868** (2017) 012020 doi:10.1088/1742-6596/868/1/012020

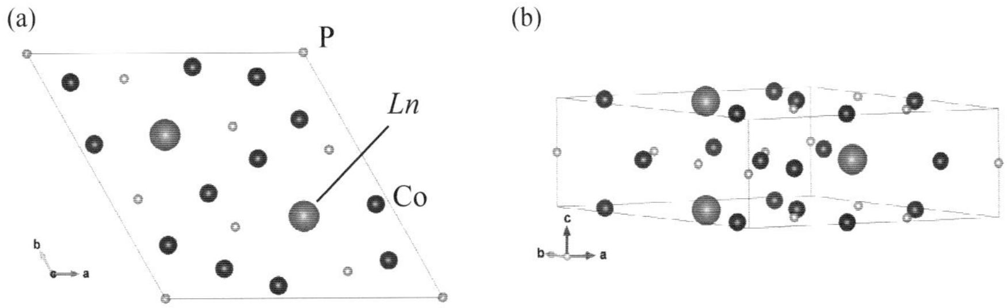

Figure 1. Crystal structure of $Ln_2Co_{12}P_7$.

[5, 6]. Especially, in the case of Ln = Nd, the spontaneous magnetization gradually decreases with the temperature below 70 K and becomes zero at the ground state [5]. Zero value of spontaneous magnetization means that $Nd_2Co_{12}P_7$ is in an antiferromagnetically ordered state. M. Reehuis *et al.* have reported that the magnetic structure at the antiferromagnetic state is that the Co sublattice moment aligns parallel to the c axis and the Nd sublattice moment aligns antiparallel to the Co sublattice moment [5]. The magnetic structure and zero spontaneous magnetization in the antiferromagnetic state indicate that the ordered moment of Nd^{3+} ion is much smaller than the value expected for an isolated Nd^{3+} ion since the Co sublattice moment is small compared with that in usual ferromagnets. To answer why the antiferromagnetic ground state realizes in $Nd_2Co_{12}P_7$, we studied the magnetic property of $Nd_2Co_{12}P_7$ through magnetic measurements and analyses by using polycrystalline and single crystal samples.

2. Experiments

We synthesized a polycrystalline sample of $Nd_2Co_{12}P_7$ by a solid-state reaction method [4]. We rst mixed powders of Nd (3N), Co (3N) and P (6N) in an atomic ratio of Nd:Co:P = 2.5:12:7 and heated this mixture in an evacuated silica tube at 1,000 °C for 10 hours. We then ground the obtained sample and heated it in evacuated silica tubes at 1,200 °C for 12 hours. We synthesized single crystals of $Nd_2Co_{12}P_7$ by a Sn flux method [5]. We heated a mixture of Nd (3N), Co (3N), P (6N) and Sn (3N) with an atomic ratio of Nd:Co:P:Sn = 3:12:7:70 in an evacuated silica tube at 880 °C for one week and quenched it. Tin-rich matrices were dissolved in slightly diluted (1:1) hydrochloric acid.

We measured magnetization (M) of samples by a vibrating sample magnetometer ($MagLab^{VSM}$, Oxford Instruments), installed in Tokyo University of Agriculture and Technology as a function of temperature (T) and magnetic eld (H). We also measured magnetization curves up to H = 56 T by using a non-destructive pulsed eld magnet, installed in Institute for Solid State Physics, The University of Tokyo.

3. Results and discussion

Figure 2 shows T dependence of inverse magnetic susceptibility (H/M) of the polycrystalline sample of $Nd_2Co_{12}P_7$ measured at $H = 0.1$ T. H/M shows a convex behavior of the paramagnetic state, indicating that magnetic susceptibility consists of two components of magnetic sublattices. To analyze H/M in the paramagnetic state, we tted the data in the range of $T > 180$ K to the following equation,

$$\frac{H}{M} = \left[12 \, \frac{N_A P_{\mathrm{eff,Co}}^2}{3k_B(T \quad \theta_{\mathrm{Co}})} + 2 \, \frac{N_A P_{\mathrm{eff,Nd}}^2}{3k_B(T \quad \theta_{\mathrm{Nd}})} \right]^{-1} \tag{1}$$

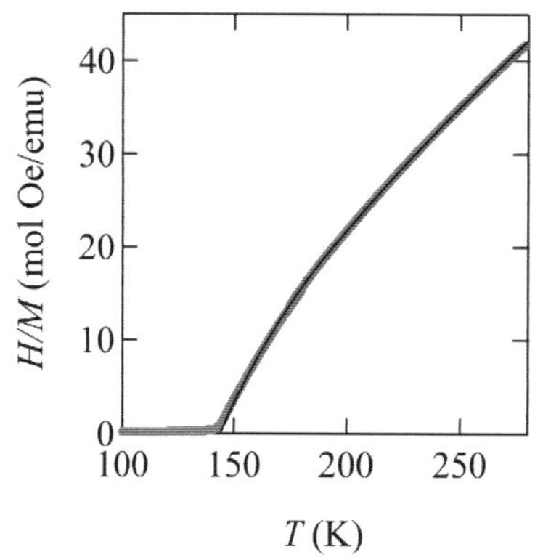

Figure 2. T dependence of H/M of the polycrysalline sample measured at 0.1 T. Black thin line is the reuslt of tting to Eq. (1).

by the least-squares method. $P_{\text{eff},X}$ is the e ective Bohr magneton ($= g_J\mu_B\sqrt{J(J+1)}$, g_J and μ_B are the Landé g-factor and the Bohr magneton, respectively) and θ_X is the Weiss temperature for X (X = Nd or Co). The k_B and N_A are the Boltzmann constant and the Avogadro constant, respectively. Here, $P_{\text{eff,Nd}}$ is held at the theoretical value of isolated Nd^{3+} ($= 3.62\ \mu_B$). As the result, we obtained values of the parameters as θ_{Nd} = - 25 K, $P_{\text{eff,Co}} = 1.09\ \mu_B$, and $\theta_{\text{Co}} = 144$ K. Since these values are almost the same with those reported in Ref. [7], we think our analysis is reasonable. Our result shows two facts. One is that the energy scale of the crystal eld that $4f$ electrons of Nd^{3+} experience is small compared with k_BT_C. The other is that the interaction between the Nd and Co sublattices is negligible in the paramagnetic state.

Figure 3 shows T dependence of M of single crystals with H applied along c-axis direction. Our results are consistent to those reported in Ref. [5]. We newly found that M shows a di erence between data measured after zero- eld cooling (ZFC) and eld cooling (FC) below 15 K. Since M-T curves measured after ZFC and FC cross at 4 K and M measured after ZFC takes negative values between 4 K and 10 K, $Nd_2Co_{12}P_7$ is in a similar state with compensated ferrimagnets in this temperature region. Although the detailed mechanism of such behavior is still unclear, it must be a key for understanding why antiferromagnetic state realizes at the ground state in $Nd_2Co_{12}P_7$.

Figure 4 shows H dependence of M of single crystals measured in steady eld up to 12 T applied along c-axis direction. In the region between 75 to 135 K, M-H curves show convex behaviors, which is typical of a ferromagnetic state. On the other hand, below 30 K, M-H curves show linear or concave behaviors, which is indicative of a ferrimagnetic state. At 2 K, there is no spontaneous magnetization, showing $Nd_2Co_{12}P_7$ is in antiferromagnetic state. These results indicate that $Nd_2Co_{12}P_7$ changes from ferromagnetic to ferrimagnetic states with the decrease in temperature and is nally in the antiferromagnetic state at the ground state, but magnetic phase transition does not exist in each change.

Figure 5 shows a magnetization curve of single crystals with applying a pulsed magnetic eld along c-axis direction at T = 4.2 K. We observed a metamagnetic transition around 15 T. Since at the zero eld, both sublattice moments of Nd and Co align with the c-axis direction and are antiparallel with each other in this T-region, this transition is understood as that the relative direction between both sublattice magnetizations changes from antiparallel to parallel with the increase in the magnetic eld. From the value of M above 30 T, we estimated the saturating magnetization of $Nd_2Co_{12}P_7$ as about 8 μ_B per formula unit.

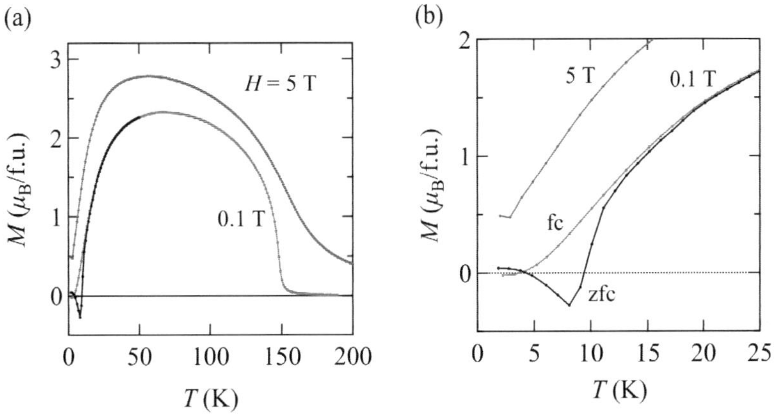

Figure 3. (a) T dependence of M of single crystals of $Nd_2Co_{12}P_7$ measured at $H = 0.1$ T and 5 T. H is applied along c-axis of single crystals. (b) Magni cation of panel (a) in the region of $T \leq 25$ K.

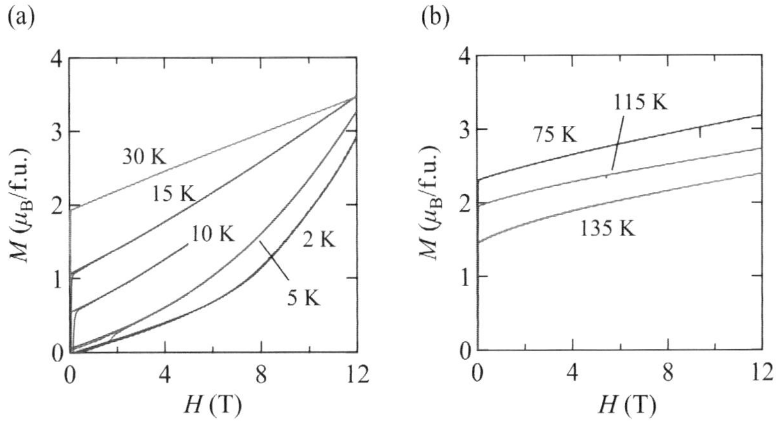

Figure 4. Magnetization curves measured at various temperatures in the T-range of (a) $T \leq 30$ K and (b) 75 K $\leq T \leq$ 135 K. H is applied along c-axis of single crystals.

We discuss the value of the Nd sublattice moment in the antiferromagnetic state. From the Fig. 4 (b), one can estimate the spontaneous magnetization at 75 K as 2.3 μ_B per formula unit. Since $Nd_2Co_{12}P_7$ is in the ferromagnetic state in this temperature region, the spontaneous magnetization must correspond to the Co sublattice moment. By a natural extrapolation to zero temperature, we roughly estimated the Co sublattice moment at the ground state as 2.5 μ_B per formula unit. In the ground state, the Nd sublattice moment is in antiparallel with the Co sublattice moment, and the value of it is the same with that of the Co sublattice moment. Therefore, the Nd sublattice moment at the ground state is 1.25 μ_B per Nd. This value is smaller than the theoretical value expected for an isolated Nd^{3+} ion (= 3.27 μ_B). We also estimated the value of the Nd sublattice moment in the high-H region. As stated above, the value of the saturating magnetization is about 8 μ_B per formula unit. In the previous report, the value of the saturating magnetization of $Lu_2Co_{12}P_7$ is 4 μ_B per formula unit [5]. If the

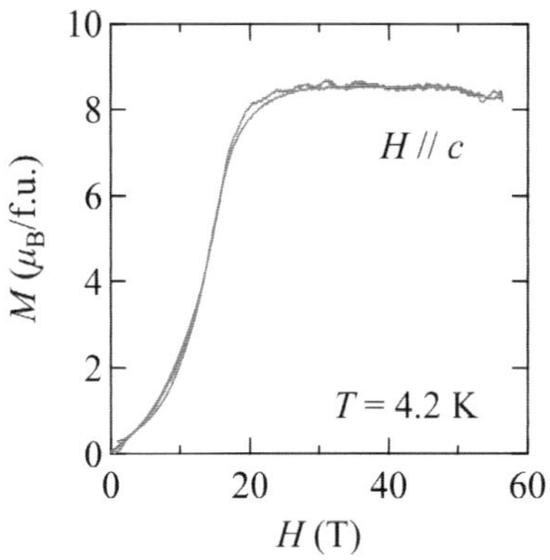

Figure 5. Magnetization curve at 4.2 K. H is applied along c-axis of single crystals.

saturating magnetization of the Co sublattice is the same among $Ln_2Co_{12}P_7$, we can estimate the Nd sublattice moment as 4 μ_B per formula unit by subtracting the value of the Co sublattice moment from the saturating magnetization. As a result, the value of the Nd sublattice moment is 2 μ_B per Nd in the high-H region. This value is larger than the value for the low-H region (1.25 μ_B per Nd) but is still smaller than the theoretical value for an isolated Nd^{3+} ion.

In the case of $Ln_2Co_{12}P_7$, the electronic state of Ln^{3+} is reported to well explained by the e ective Hamiltonian $\hat{H}_{\rm eff}$ as described by the following equation [6],

$$\hat{H}_{\rm eff} = \hat{H}_{\rm cfe} + \lambda\vec{S}\ \ \vec{L} + \ \ \vec{H}_{\rm ex}\ \ \vec{S}, \quad (2)$$

where the rst and second terms correspond to crystal eld e ect and spin-orbital coupling, respectively. The third term corresponds to coupling between the spin part of $4f$ electrons and the exchange eld ($H_{\rm ex}$) derived from the ferromagnetic sublattice. The λ is the spin-orbital coupling constant. It has been succeeded in explaining a electronic state of Ln^{3+} in the base materials of permanent magnet by Eq. (2) [8]. Here, the constatn is equal to +2 in case of permanent magnet [8], and is less than zero in case of $Ln_2Co_{12}P_7$ [6]. Therefore, the spin part of magnetic moment of Ln^{3+} tends to align in parallel with $H_{\rm ex}$ in the case of $Ln_2Co_{12}P_7$. As a result, the total magnetic moment of Nd ions aligns in antiparallel with the ferromagnetic Co sublattice moment since $\lambda > 0$ in the case of Nd. Figure 6 schematically shows how magnetic moments are coupled each other. In the paramagnetic state, $4f$ electrons of Nd^{3+} are expected to be in the same state with the isolated one, whose magnetic moment is $g_J J = 3.27$ μ_B. However, below 5 K, the Nd sublattice moment is smaller than the expected value from the value of $g_J J$ of an isolated Nd^{3+} ion in both the low-H and high-H regions. Also, the temperature dependence of the Nd sublattice moment is anomalous as seen in Fig. 3 (a). These unusual behaviors can be understood by the model as following. In the Nd sublattice, where each Nd^{3+} has the same value of magnetic moment with that of isolated Nd^{3+} ions, ordered moments on Nd sites are gradually induced by $H_{\rm ex}$ of the Co sublattice with decreasing temperature. The induced moments of Nd^{3+} ions nally cancel the Co sublattice moment below 5 K. For now, we do not know whether there is a necessity for the complete cancellation of sublattice moments at the ground state or not. It is also possible that magnetic moments of Nd^{3+} ions take a smaller value in the low-T region compared with the value in the paramagnetic state due to the crystal eld splitting of the lowest energy term $^4I_{9/2}$. To understand anomalous magnetism of $Nd_2Co_{12}P_7$, further studies are needed.

IOP Conf. Series: Journal of Physics: Conf. Series **868** (2017) 012020 doi:10.1088/1742-6596/868/1/012020

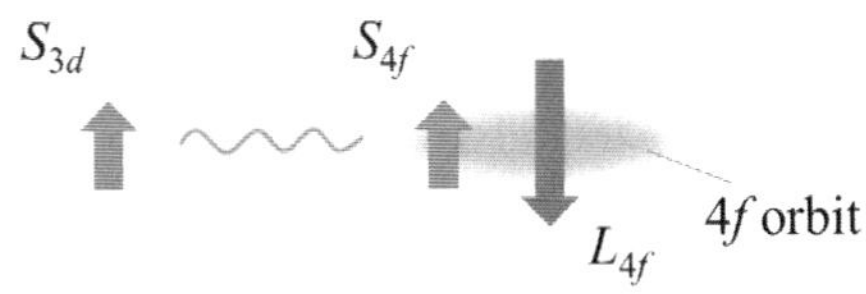

Figure 6. Schematic explanation of interaction between the ferromagnetic Co sublattice moment (S_{3d}) and the spin part of the magnetic moment (S_{4f}) and the orbital part of the magnetic moment (L_{4f}) of Nd^{3+}.

4. Conclusion

We synthesized polycrystalline and single crystal samples of $Nd_2Co_{12}P_7$ and measured magnetization of the samples. We observed a gradual increase of the Nd sublattice moment with decreasing temperature, which is quite di erent from the temperature dependence of ordered moments in an ordinary two-sublattice model. Also, we newly observed anomalous hysteresis in the temperature dependence of magnetization below 15 K which reminds us compensated ferrimagnets, although $Nd_2Co_{12}P_7$ is in the antiferromagnetic state in the ground state. In the antiferromagnetic state, we con rmed that the Nd sublattice moment takes a smaller value than the value expected for isolated Nd^{3+} ions. We suggested that anomalous behavior of the Nd sublattice moment is due to the coupling between localized magnetic moments of Nd ions and a ferromagnetic exchange eld.

Acknowledgments

This work was supported by Grants-in-Aid for Young Scientists (B) from Japan Society for the Promotion of Science (Grant Nos. 24760534 and 15K18211).

References

[1] Ohta H and Yoshimura K 2009 *Phys. Rev. B* **80** 184409
[2] Ohta H, Michioka C and Yoshimura K 2011 *Phys. Rev. B* **84** 134411
[3] McGuire M A, Gout D J, Garlea V O, Sefat A S, Sales B C and Mandrus D 2010 *Phys. Rev. B* **81** 104405
[4] Jeitschko W, Braun D J, Ashcraft R H and Marchand R 1978 *J. Solid State Chem.* **25** 309
[5] Reehuis M, Ouladdiaf B, Jeitschko W, Vomhof T, Zimmer B and Ressouche E 1997 *J. Alloys Comp.* **261** 1
[6] Ohta H, Watanabe Y, Miyake A, Tokunaga M and Aruga Katori H 2016 *J. Jpn. Soc. Powder Powder Metallurgy* **63** 652
[7] Reehuis M and Jeitschko W 1989 *J. Phys. Solids* **50** 563
[8] Franse J J M and Randwański R J 1993 *Handbook of Magnetic Materials* **7** 307

Paramagnetic-to-nonmagnetic transition in antiperovskite nitride Cr_3GeN studied by ^{14}N-NMR and μSR

K Takao[1], Z Liu[1], K Uji[1], T Waki[1], Y Tabata[1], I Watanabe[2] and H Nakamura[1]

[1] Department of Materials Science and Engineering, Kyoto University, Kyoto 606-8501, Japan
[2] Advanced Meson Science Laboratory, RIKEN Nishina Center, Wako 351-0198, Japan

E-mail: takao.kenta.42w@st.kyoto-u.ac.jp

Abstract. The antiperovskite-related nitride Cr_3GeN forms a tetragonal structure with the space group $P\bar{4}2_1m$ at room temperature. It shows a tetragonal ($P\bar{4}2_1m$) to tetragonal ($I4/mcm$) structural transition with a large hysteresis at 300–400 K. The magnetic susceptibility of Cr_3GeN shows Curie-Weiss type temperature dependence at high temperature, but is almost temperature-independent below room temperature. We carried out μSR and ^{14}N-NMR microscopy measurements to reveal the magnetic ground state of Cr_3GeN. Gradual muon spin relaxation, which is nearly temperature-independent below room temperature, was observed, indicating that Cr_3GeN is magnetically inactive. In the ^{14}N-NMR measurement, a quadrupole-split spectrum was obtained at around $^{14}K = 0$. The temperature dependence of $^{14}(1/T_1)$ satisfies the Korringa relation. These experimental results indicate that the ground state of Cr_3GeN is Pauli paramagnetic, without antiferromagnetic long-range order.

1. Introduction

The antiperovskite compounds ($M_3M'X$, M = 3d transition metal or rare earth, M' = metal or metalloid, X = B, C, N) have been invesestigated for several decades, and various kinds of elements with different M and M' have been discovered [1, 2, 3, 4, 5, 6]. They form a cubic structure with the space group $Pm\bar{3}m$; M' atoms are located in the corner positions, while M and X atoms occupy the face-centered and body-centered positions, respectively. The magnetic properties of Mn-based compounds have been widely studied with respect to their application in electrical technologies [7, 8, 9]. However, few studies have investigated the magnetism of Cr-based compounds.

The Cr-based compound Cr_3GeN forms an antiperovskite-related structure. Its magnetic susceptibility has been reported [2, 10]. The structure of Cr_3GeN at room temperature is tetragonal with the space group $P\bar{4}2_1m$ [1] and transforms to another tetragonal structure with $I4/mcm$ at higher temperatures. In addition, there is a large hysteresis at the $P\bar{4}2_1m$ to $I4/mcm$ transition at temperatures in the range 300–400 K. The susceptibility shows a Curie-Weiss behavior in the high temperature phase. Below the transition temperature T^*, the susceptibility exhibits a small value down to 5 K [10]. The previous magnetic susceptibility measurement suggested an antiferromagnetic state as the low temperature phase [10]. However, there is a lack of microscopic information on the magnetic state of Cr.

In the present study, we investigated the magnetic ground state of Cr_3GeN using nuclear magnetic resonance (NMR) and muon spin relaxation (μSR) microscopy measurements, which reveal that Cr loses its moment below T^*, indicating that the magnetic ground state is not antiferromagnetic but nonmagnetic.

2. Experimental

A polycrystalline sample of Cr_3GeN was prepared by a solid-state reaction using Cr_2N, Cr and Ge as the starting materials. They were mixed with a stoichiometrical ratio, pressed into a pellet, sealed in an evacuated quartz tube and then calcined at 1173 K for 5 days. The sample was characterized by room-temperature powder X-ray diffraction (XRD) measurement (X'Pert PRO Alpha-1, PANalytical). The sample primarily consisted of Cr_3GeN, although small amounts of impurities (1.5 % Cr_2O_3 and 1.1 % GeO_2) were found, which do not substantially affect on our conclusion. The temperature dependence of magnetic susceptibility was measured at 2–1120 K using a SQUID magnetometer (MPMS, Quantum Design) and VSM by applying a magnetic field of up to 0.5 T. NMR measurements were carried out using a phase-coherent spin-echo pulse spectrometer. The ^{14}N-NMR ($I = 1$, $\gamma/2\pi = 3.075$ MHz/T) was measured in an external magnetic field of 6.19 T. The NMR spectrum at 4.2 K was taken by summing the Fourier transform of the spin-echo signals at various frequencies. The spin-lattice relaxation time was obtained by the saturation-recovery method. Zero-field (ZF) μSR measurements were performed at the RIKEN-RAL Muon Facility at the Rutherford-Appleton Laboratory in the UK using a pulsed positive surface muon beam at 11.6–350 K.

3. Results and Discussion

Figure 1 shows the temperature dependence of the magnetic susceptibility. The result below 700 K corresponds well with data from the literature [2, 10]. The anomalies at 850 and 950 K correspond to other structural transitions [11].

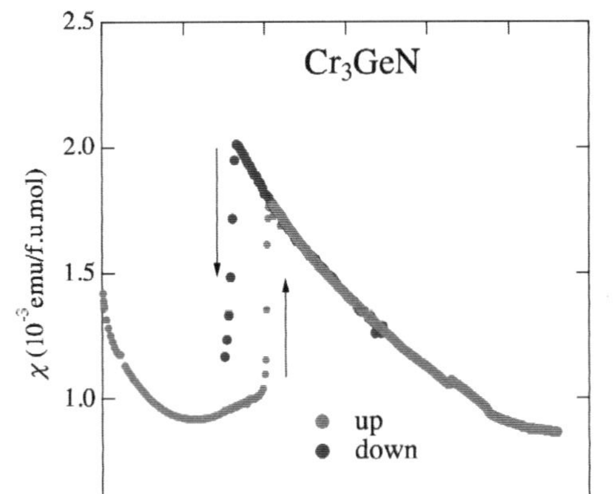

Figure 1. The temperature dependence of magnetic susceptibility for Cr_3GeN at 2–1120 K.

We performed microscopic measurements to demonstrate the disappearance of Cr moment in the ground state. Figure 2 shows a ^{14}N-NMR quadrupole-split spectrum. The spectrum has a characteristic double peak structure and is well fitted by a simulated powder pattern for the ^{14}N nucleus with $I = 1$ and the asymmetric parameter $\eta = 0$, since the nitrogen is located at an axially symmetric site. Both the quadrupole frequency, ν_Q, and the shift, K, are small; $\nu_Q = 0.27$ MHz and $K = -0.23\%$. The small K indicates no internal field in the ground state.

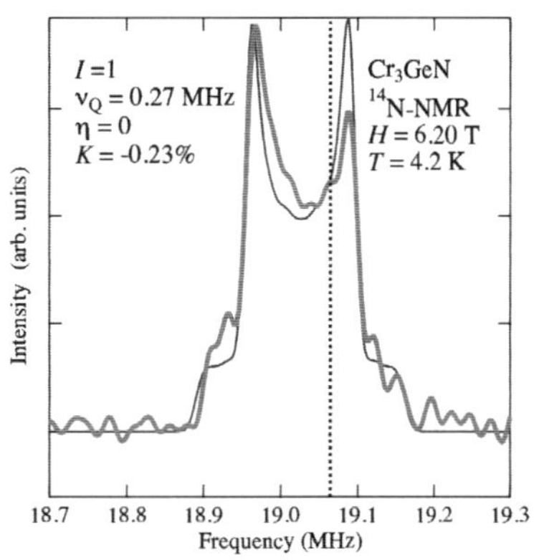

Figure 2. The Fourier transform ^{14}N NMR spectrum of Cr_3GeN at 4.2 K. The red curve shows the experimental data, and blue curve is a simulation. The dotted line represents the frequency at $K = 0$.

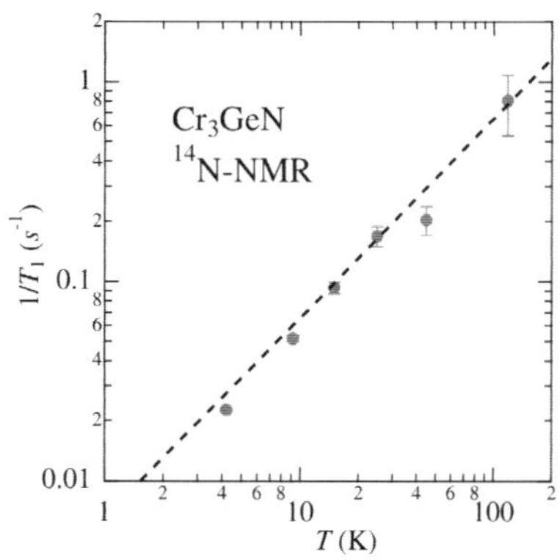

Figure 3. The temperature dependence of the ^{14}N spin-lattice relaxation rate for Cr_3GeN. The dashed line denotes the fit of experimental data with the Korringa law.

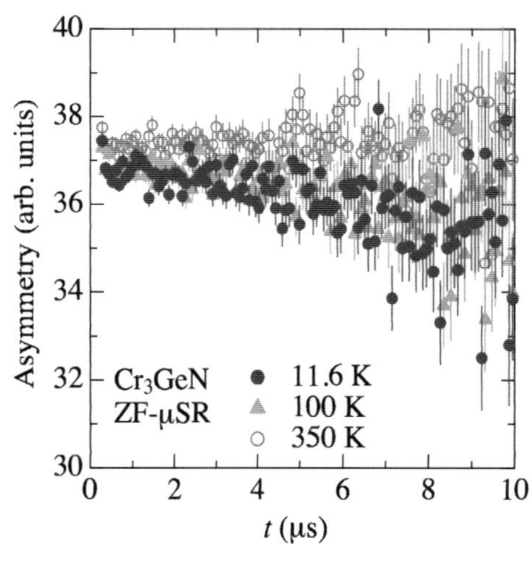

Figure 4. ZF-μSR spectra of Cr_3GeN measured at 11.6, 100 and 350 K.

^{14}N spin-lattice relaxation measurements were performed at various temperatures up to 120 K. The spin-lattice relaxation time, T_1, was estimated by single exponential fitting. Figure 3 shows the temperature dependence of $1/T_1$, which is proportional to T. The Korringa behavior indicates that the relaxation is dominated by conduction electrons. Moreover, T_1 is very long, say, 44 s at 4.2 K suggesting that the density of states at the Fermi level is markedly reduced at low temperature.

ZF-μSR measurements give useful information on the presence of the magnetic moment and magnetic ordering. Figure 4 shows ZF-μSR spectra. At all temperatures, the muon spin asymmetry does not show any precession. This means that there is no magnetic ordering. In addition, the relaxation is very slow, which is consistent with the fact that the Cr magnetic moment disappears below T^*.

4. Conclusion

We investigated the magnetic ground state of Cr_3GeN by using microscopic methods, ^{14}N-NMR and μSR, which demonstrated that the compound is the Pauli paramagnetic. ZF-μSR measurements suggest the disappearance of Cr moment below the structural transition, which is different from the antiferromagnetism as expected. The origin of the nonmagnetic ground state is still open to question. Band calculation would help reveal the mechanism of the disappearance of Cr moment at the structural transition.

References

[1] Boller H 1969 *Monatsh. Chem.* **100** 1471
[2] Nardin M, Lorthioir G, Barberon M, Madar R, Fruchart E and Fruchart R 1972 *C. R. Acad. Sci. C Chim.* **274** 2168
[3] Boller H and Nowotny H 1968 *Monatsh. Chem.* **99** 721
[4] Jeitschko W, Nowotny H and Benesovsky F 1964 *Monatsh. Chem.* **95** 436
[5] Samson C, Bouchaud J P and Fruchart R 1964 *C. R. Hebd. Seances Acad. Sci.* **259** 392
[6] Haschke H, Nowotny H and Benesovsky F 1966 *Monatsh. Chem.* **97** 716
[7] Kodama K, Iikubo S, Takenaka K, Takigawa M, Takagi H and Shamoto S 2010 *Phys. Rev. B* **81** 224419
[8] Kamishima K, Goto T, Nakagawa H, Miura N, Ohashi M, Mori N, Sasaki T and Kanomata T 2000 *Phys. Rev. B* **63** 024426
[9] Takenaka K and Takagi H 2005 *Appl. Phys. Lett.* **87** 261902
[10] Lin S, Tong P, Wang B S, Huang Y N, Song W H and Sun Y P 2014 *J. Alloys Compd.* **584** 308
[11] Wang B and Ohgushi K 2016 *Sci. Rep.* **6** 37896

Charge fluctuation in $S = 1/2$ triangular lattice cluster antiferromagnets $Li_2ScMo_3O_8$ and $Li_2InMo_3O_8$

Yuya Haraguchi[1], **Chishiro Michioka**[1], **Hiroaki Ueda**[1], **Kazuyoshi Yoshimura**[1,2]

[1] Department of Chemistry, Graduate School of Science, Kyoto University, Kyoto 606-8502, Japan

[2] Research Center for Low Temperature and Materials Sciences, Kyoto University, Kyoto 606-8501, Japan

E-mail: `chiyuya@kuchem.kyoto-u.ac.jp`, `kyhv@kuchem.kyoto-u.ac.jp`

Abstract. We investigated macroscopic and microscopic physical properties of $S = 1/2$ Mo_3 cluster magnets $Li_2ScMo_3O_8$ and $Li_2InMo_3O_8$. $Li_2InMo_3O_8$ shows magnetic ordering at $T_N \simeq 12$ K, while no magnetic ordering is observed down to 0.5 K in $Li_2ScMo_3O_8$ in spite of the strong antiferromagnetic interaction among clusters probed by the Weiss temperature. ^{115}In and ^{45}Sc NMR measurements demonstrate that there should be difference in charge fluctuation in clusters between $Li_2InMo_3O_8$ and $Li_2ScMo_3O_8$. Our result suggests that the origin of different ground states realized in $Li_2InMo_3O_8$ and $Li_2ScMo_3O_8$ is different condition in charge and local structure as one of characteristics of the cluster magnet.

1. Introduction

In condensed matter physics, spin frustrated magnets have attracted much interest for their ground states and low energy excitations[1, 2, 3]. In theoretical predictions, the spin frustration effect leads to the emergence of an exotic ground state such as a resonating valence bond state, a spin liquid state, and so on. A strong spin frustration is realized in a characteristic lattice like triangular, kagome (in 2D), face-centered cubic and pyrochlore lattice (in 3D). With this kind of background, many challenges of searching model compounds to study spin frustration have been accomplished.

Recently, we newly discovered $S = 1/2$ triangular lattice cluster antiferromagnet $Li_2ScMo_3O_8$ which has the isostructure of $Li_2InMo_3O_8$ [4]. Their ground states are completely different from each other; while $Li_2InMo_3O_8$ exhibits a conventional magnetic ordering with 120° structure, $Li_2ScMo_3O_8$ show a spin liquid like condensation [4]. In the theoretical predictions, charge and lattice fluctuation as characteristics of the cluster based triangular lattice compound would lead their ground state to spin liquid state [5, 6]. Therefore, it is necessary to perform their microscopic structural studies to reveal the origin of different ground states of $Li_2ScMo_3O_8$ and $Li_2ScMo_3O_8$.

In this work, we report on ^{45}Sc/^{115}In nuclear magnetic resonance (NMR) measurements for $Li_2ScMo_3O_8$ and $Li_2InMo_3O_8$. From the ^{45}Sc/^{115}In-NMR studies, charge fluctuations are found to play a key role for suppression of a magnetic ordering in $Li_2ScMo_3O_8$. $Li_2ScMo_3O_8$ is found to be a novel spin liquid candidate produced in the cooperating system between spin and charge fluctuations.

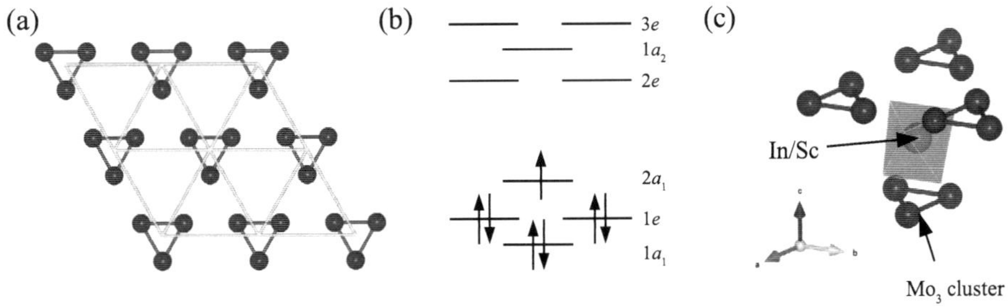

Figure 1. (Color online) (a) $[Mo_3]^{11+}$ trimers forming the triangular lattice. (b) Schematic 4d orbital energy level of the Mo_3 trimer with the electron configurations expected in $Li_2AMo_3O_8$ (A = In or Sc). (c) Coordination of Mo_3 clusters around In/Sc ion.

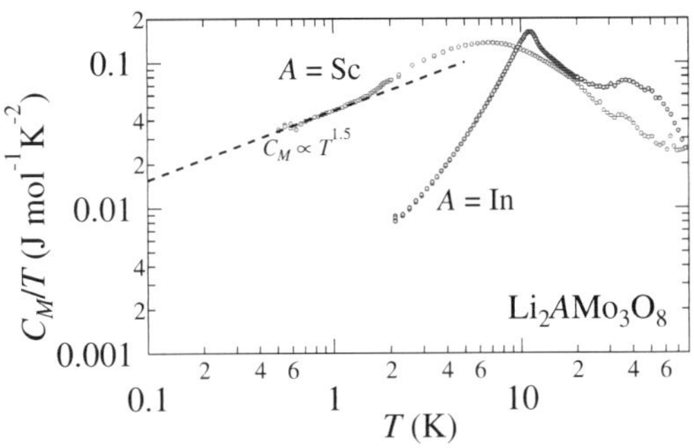

Figure 2. (Color online) Temperature dependence of the magnetic heat capacity divided by temperature C_M/T in $Li_2InMo_3O_8$ and $Li_2ScMo_3O_8$ reproduced from ref. [4]. The dashed line is fit to the power law ($C_M \propto T^{\alpha}$) with $\alpha = 1.5$ for $Li_2ScMo_3O_8$.

2. Experimental Methods

Polycrystalline samples of $Li_2ScMo_3O_8$ and $Li_2InMo_3O_8$ were prepared using conventional solid-state reactions as describe in a previous report [4]. The thermodynamic properties were measured by using a conventional relaxation method with a physical property measurement system (PPMS; Quantum Design). NMR measurements were performed by the spin-echo method by using a standard phase coherent pulse spectrometer. The ^{115}In and ^{45}Sc nuclei have nuclear spins $I = 9/2$ and $7/2$ and gyromagnetic ratios $^{115}\gamma = 9.3301$ MHz/kOe and $^{45}\gamma = 10.343$ MHz/kOe, respectively.

3. Results and Discussion

As shown in the inset of Fig. 2, the temperature derivative $d\chi/dT$ for $Li_2InMo_3O_8$ decreases discontinuously at $\sim$ 12 K indicating the presence of long-range ordering, which is also suggested by the result of heat capacity measurement as mentioned below. In contrast, $d\chi/dT$ for $Li_2ScMo_3O_8$ does not show any discontinuous changes down to 2 K. The specific heat data provides further evidence of the formation of the spin-liquid state. The magnetic specific heat C_M is estimated by subtracting the specific heat of isomorphic nonmagnetic compound $Li_2SnMo_3O_8$ as a lattice contribution. Figure 2 shows the temperature dependence of the

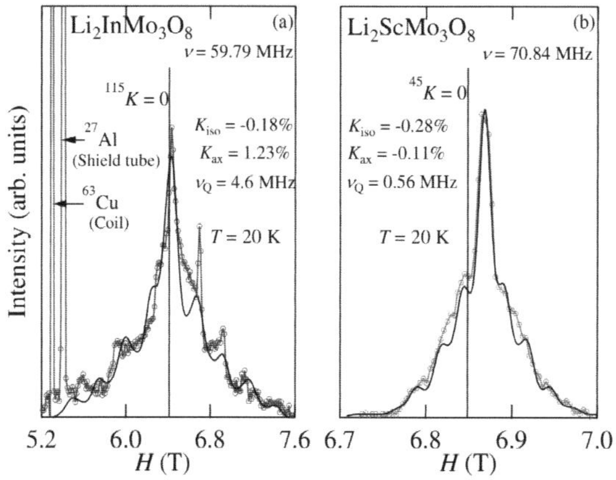

Figure 3. (Color online) Magnetic field-swept ^{115}In and ^{45}Sc NMR spectra at 20 K for (a) $Li_2InMo_3O_8$ and (b) $Li_2ScMo_3O_8$ with constant frequencies νs of 59.79 and 70.84 MHz for $Li_2InMo_3O_8$ and $Li_2ScMo_3O_8$, respectively. Simulated powder-pattern spectra are also presented by solid lines for comparison . The simulated conditions of the isotropic and anisotropic shifts $K_{\rm iso}$ and $K_{\rm ax}$, and the pure nuclear quadrupole frequency $\nu_{\rm Q}$ are written in the figure for each nuclei.

magnetic specific heat divided by temperature (C_M/T) for $Li_2InMo_3O_8$ and $Li_2ScMo_3O_8$. In the case of $Li_2InMo_3O_8$, the λ-shaped peak was observed at the same temperature, suggesting the existence of long-range ordering. In the case of $Li_2ScMo_3O_8$, we observed no λ-typed anomaly in C_M/T down to 0.5 K but only a broad hump near 10 K, indicating the absence of any magnetic orderings. The C_M of $Li_2ScMo_3O_8$ between 0.5 and 4 K are well fitted by the power-law $C_M \simeq T^{1.5}$. Such power low behavior of C_M suggests the presence of gapless excitations in low temperatures.

In the theoretical predictions, the ground state of the cluster-based triangular lattice systems is ruled by the charge and lattice fluctuations within a cluster [5, 6]. Thus, the origin of the different ground states between $Li_2ScMo_3O_8$ and $Li_2InMo_3O_8$ could be the difference in the local structure around a Mo_3O_{13} cluster. To clarify the difference of the local structure, we have measured the ^{45}Sc and ^{115}In NMR spectra for $Li_2ScMo_3O_8$ and $Li_2InMo_3O_8$, respectively at 20 K sweeping a magnetic field at constant frequencies described in the figure. These results are plotted in Fig. 4. The observed spectra are well reproduced by powder-pattern simulations with second-order quadrupole effects using the three parameters listed in the figure, suggesting that In and Sc ions occupy only single site in both compounds. The electric field gradient at Sc and In sites can be estimated with using the values of pure nuclear quadrupole frequency ν_Q derived from the following function,

$$\nu_Q = \frac{3eQ \cdot (eq)}{2hI(2I-1)}, \tag{1}$$

where e is the elementary charge, h is the Planck's constant, q is the electric quadrupole moment, Q is the nuclear electric quadrupole moment ($Q = -0.22 \times 10^{-24}$cm^2 for Sc and 1.66×10^{-24}cm^2 for In), and I is the nuclear spin quantum number ($I = 7/2$ for Sc and $9/2$ for In). Estimated values of eq are 3.89×10^{-17} and 1.47×10^{-17} J·C^{-1}cm^{-2} for In of $Li_2InMo_3O_8$ and Sc of $Li_2ScMo_3O_8$, respectively. The quadrupole asymmetry parameter η is zero because that the In/Sc site is on a hexagonal axis. The value of the electric field gradient eq is expected to be contributed only by environmental ions owing to the closed-shelled In^{3+} and Sc^{3+} ions. In and Sc sites occupy almost the same environmental site; hence, the relation of $eq(In^{3+}) \simeq eq(Sc^{3+})$ can be expected. However, eq for In^{3+} is 2.6 times larger than that for Sc^{3+}, indicating a strong suppression of eq at Sc site. The difference of eq is difficult to be explained by a simple point charge model because of the similar environmental In and Sc sites.

The shielding effect derived from the charge fluctuations in the clusters can explain the origin of large difference in eq for In^{3+} of $Li_2InMo_3O_8$ and Sc^{3+} of $Li_2ScMo_3O_8$. As shown in Fig. 1(c), Sc/In ions surrounded by the Mo_3 cluster. Thus, Sc/In sites are thought to be susceptible to the charge fluctuation within the cluster. Chen *et al.* expressed the occurrence in the spin-disordered ground state of Mo_3 cluster magnet by using an extend Hubbard model taking into account of electron transfers of intra- and inter-cluster [6]. When both inter- and intra-cluster transfers are effective, the electron charge should fluctuate without strong localization. Such the charge fluctuation explaining the shielding effect of eq possibly suppresses the magnetic ordering. There is a possibility that spins in the intermediate charge glass state undergo a dynamical slowing down.

In the scenario to explain the difference of eq, the charge fluctuations suppress the magnetic ordering in $Li_2ScMo_3O_8$, which is in contrast to the case of $Li_2InMo_3O_8$ with 120° structural magnetic ordering. In $Li_2ScMo_3O_8$, the observed power law behavior of $C_m \simeq T^{1.5}$ suggests a gapless quantum spin liquid in the low temperatures. In a similar case of $S = 1/2$ triangular lattice cluster compound, κ-(BEDT-TTF)$_2$Cu$_2$(CN)$_3$ exhibits power law behavior in the temperature dependence of the nuclear spin-lattice relaxation rate $(1/T_1)$[8, 9]. On the other hand, the result of thermal-transport measurement supports the existence of a gapped quantum spin liquid phase[10]. In other cases, spin liquid candidates of triangular lattice Heisenberg antiferromagnets $NiGa_2S_4$[11, 12] and $Ba_3IrTi_2O_9$[13] exhibit the power law behaviors as the power index $\alpha \simeq 2$ in $C_m = T^{\alpha}$. In the spin liquid ground state on the triangular lattice Heisenberg antiferromagnet, the presence/absence of the spin gap in the low energy spin excitations have been of great interest. Our results of $Li_2ScMo_3O_8$ indicate that there is gapless excitations in the low temperature regime, although the observed power index is somewhat different from the above mentioned gapless spin liquid candidates.

4. Summary

We studied the microscopic physical properties of $Li_2ScMo_3O_8$ and $Li_2InMo_3O_8$. These compounds are categorized in the triangular lattice antiferromagnetic cluster magnet. $Li_2InMo_3O_8$ shows the magnetic ordering at $T_N \simeq 12$ K, while $Li_2ScMo_3O_8$ shows no magnetic ordering down to 0.5 K with gapless excitations in the low temperature regime. From ^{115}In and ^{45}Sc NMR measurements, it has been revealed that there is the difference in eq at crystallographically equivalent In and Sc sites. This fact suggests the effective charge fluctuation stabilize the spin liquid state rather than the magnetic ordering state. Such fluctuations as a characteristic of cluster magnets are regarded as an important factor in these compounds.

Acknowledgments

This work was supported by Grants-in-Aid for Scientific Research from the Japan Society for Promotion of Science (26410089).

References

[1] Anderson P W 1973 *Mat. Res. Bull.* **8**, 153
[2] Moessner R and Sondhi S L 2002 *Prog. Theor. Phys. Suppl.* **145**, 37
[3] Lee P A 2008 *Science* **321**, 1306
[4] Haraguchi Y, Michioka C, Imai M, Ueda H and Yoshimura K 2015 *Phys. Rev. B* **92**, 014409
[5] Flint R and Lee P A 2013 *Phys. Rev. Lett.* **111**, 217201
[6] Chen G, Kee H Y and Kim Y B 2014 arXiv:cond-mat/14081963

IOP Conf. Series: Journal of Physics: Conf. Series **868** (2017) 012022 doi:10.1088/1742-6596/868/1/012022

[7] Gall P, Orabi R A R A, Guizouarn T and Gougeon P 2013 *J. Solid State Chem.* **208**, 99-102
[8] Shimizu Y, Miyagawa K, Kanoda K, Maesato M and Saito G 2003 *Phys. Rev. Lett.* **91**, 107001
[9] Yamashita S, Nakazawa Y, Oguni M, Oshima Y, Nojiri H, Shimizu Y, Miyagawa K and Kanoda K 2008 *Nat. Phys.* **4**, 459
[10] Yamashita M, Nakata N, Kasahara Y, Sasaki T, Yoneyama N, Kobayashi N, Fujimoto S, Shibauchi T and Matsuda Y 2009 *Nat. Phys.* **5**, 44
[11] Nakatsuji S, Nambu Y, Tonomura H, Sakai O, Jonas S, Broholm C, Tsunetsugu H, Qiu Y and Maeno Y 2005 *Science* **309**, 1697
[12] Nakatsuji S, Nambu Y, Onuma K, Jonas S, Broholm C and Maeno Y 2007 *J. Phys.: Condens. Matter* **19**, 145232
[13] Dey T, Mahajan A V, Khuntia P, Baenitz M, Koteswararao B and Chou F C 2012 *Phys. Rev. B* **86**, 140405

IOP Conf. Series: Journal of Physics: Conf. Series **868** (2017) 012023 doi:10.1088/1742-6596/868/1/012023

NMR studies on $EuNi_2Si_2$ with trivalent Eu ion

T Koyama[1], F Ueyama[1], T Maruyama[1], K Ueda[1], T Mito[1], A Mitsuda[2] and H Wada[2]

[1]Graduate School of Material Science, University of Hyogo, Kamigori, Hyogo 678-1297, Japan
[2]Department of Physics, Kyushu University, 744 Motooka, Nishi-ku, Fukuoka 819-0395, Japan

E-mail: t-koyama@sci.u-hyogo.ac.jp

Abstract. We have performed ^{29}Si nuclear magnetic resonance experiment to investigate microscopic magnetic properties of $EuNi_2Si_2$ with trivalent Eu ion. The temperature dependence of the nuclear spin-lattice relaxation rate $1/T_1$ can be explained by considering the two relaxation processes: conduction electrons and the thermally activated spins in the excited levels of J-multiplet. The temperature dependence of the Knight shift could roughly be reproduced with the spin-orbit coupling constant estimated from $1/T_1$.

1. Introduction

Magnetism in Europium (Eu) compounds differs substantially depending on the electronic configuration of Eu ion. According to Hund's rule, the divalent Eu ions ($4f^7$) bear a large $4f$ magnetic moment (J=S=7/2 and L=0) and its magnetic susceptibility χ obeys the Curie-Weiss law. On the other hand, the trivalent Eu ions ($4f^6$) is nonmagnetic ground state ($L = S = 3$, J=0). Here, $S(L)$ is the total spin (orbital) angular momentum and J is the total angular momentum. One of the characteristics of trivalent Eu ion is the narrow energy separation between the first-excited J-multiplet $^7F_1$ and the ground-state $^7F_0$. In the Russell-Saunders coupling scheme, the energy difference between adjacent levels is determined by the spin-orbit coupling constant λ and be expressed by

$$E_J - E_{J-1} = \lambda J. \tag{1}$$

As a result, the energy separation between $^7F_0$ and the first excited level is of the same order as k_BT for Eu^{3+} and Sm^{2+}. Therefore, the magnetism of the J-multiplet level is expected to contributes to χ [1].

In most of the intermetallic Eu compounds, the Eu^{2+} state rather than the Eu^{3+} state tends to be stable. Then only a few intermetallic Eu-based compounds have been reported to be in the Eu^{3+} state, such as $EuPd_3$ [2, 3], $EuNi_5$ [4], and $EuCo_2Si_2$ [5].

$EuNi_2Si_2$ crystallizes in the $ThCr_2Si_2$ tetragonal structure and is also categorized as the Eu^{3+} system. The trivalent Eu ions are identified on the basis of the Eu L_{III} x-ray-absorption spectroscopy and the Eu Mössbauer effect [6]. However, magnetic properties of $EuNi_2Si_2$ have not been reported. We performed the ^{29}Si nuclear magnetic resonance (NMR) measurement to investigate microscopically the magnetic properties and discussed a contribution of J-multiplet levels to static and dynamical properties of $EuNi_2Si_2$.

IOP Conf. Series: Journal of Physics: Conf. Series **868** (2017) 012023 doi:10.1088/1742-6596/868/1/012023

2. Experimental

A polycrystalline sample of $EuNi_2Si_2$ was prepared by melting mixture of a Eu ingot and NiSi alloys in a molar rate of Eu : NiSi=1.1 : 2 in an argon arc furnace to compensate the loss of Eu during melting. The x-ray diffraction pattern shows that the formation of tetragonal $EuNi_2Si_2$ with a trace of the secondary phase. The sample shows the upturn in χ at low temperatures, implying a possible inclusion of a material with magnetic Eu^{2+}. For the NMR experiments, the sample was crushed into powder. The NMR measurement was carried out using a phase-coherent spin-echo spectrometer at the temperature range of 1.5 -240 K and a constant magnetic field of 6.01 T. The NMR spectra were taken by changing the frequency and recording the integrated spin-echo intensity. The nuclear-spin lattice relaxation rate $1/T_1$ was measured using the saturation recovery method. The nuclear magnetization recovery curves could be fitted at any temperature with a standard single exponential function expected for these $I=1/2$ nuclei.

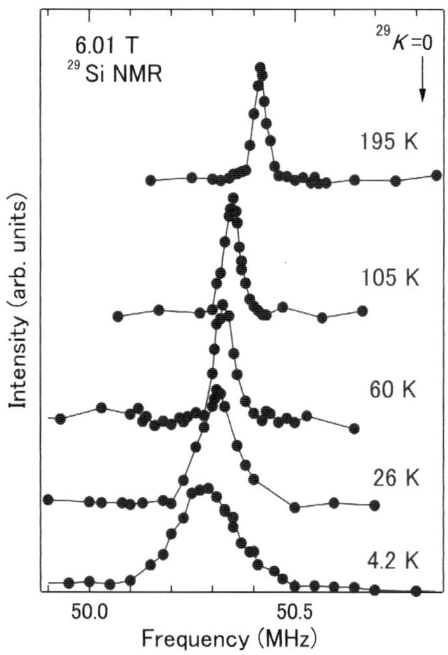

Figure 1. Temperature dependence of ^{29}Si NMR spectra.

3. Experimental results

Figure 1 shows the temperature dependence of the ^{29}Si NMR spectra. With increasing temperature, the spectra shift to higher frequency and decreases in width. Figure 2 shows the temperature dependence of Knight shift $K(T)$ derived from the peak position of the spectra. $K(T)$ shows negative values and an increase with temperature, following the convex downward curvature above 50 K. This will be discussed in the next section.

Figure 3 shows the temperature dependence of $1/T_1$ measured at the peak of the ^{29}Si NMR spectra. At low temperatures, $1/T_1$ increases proportional to temperature as is typical for nonmagnetic metals, where the relaxation process is dominated by the thermal excitations of conduction electrons at the Fermi level. With further warming, $1/T_1$ increases more rapid, suggesting the appearance of an additional relaxation channel.

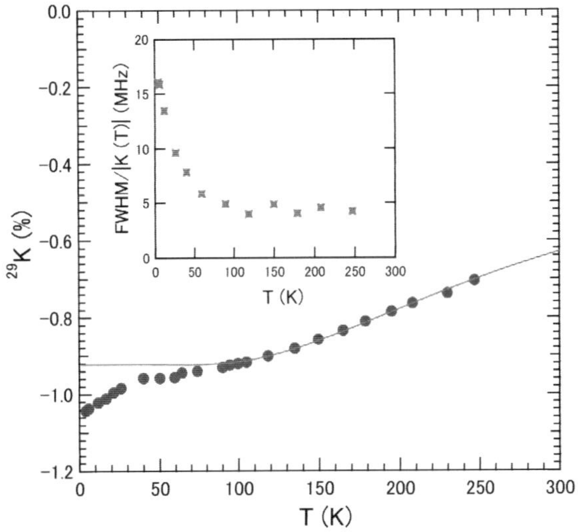

Figure 2. Temperature dependence of the Knight shift. The solid red line represents the result of a calculation based on the Russell-Saunders coupling scheme for λ=473 K deduced from the $1/T_1$ (see text). The inset shows FWHM divided by $|K(T)|$ vs temperature.

4. Discussion

First we will discuss the contribution of the excited J-multiplet levels to $1/T_1$. We assume that the observed $1/T_1$ is decomposed into two contributions as

$$\frac{1}{T_1} = \left(\frac{1}{T_1}\right)_{\mathrm{c}} + \left(\frac{1}{T_1}\right)_{\mathrm{exc}}, \tag{2}$$

where the first term in the right side is the contribution of conduction band which follows the Korringa law ($1/T_1 \propto T$). The second term is the thermally activated 4f spins in the excited states of J-multiplet. Because of the nonmagnetic ground state of $^7F_0$ and the narrow energy separation between $^7F_0$ and magnetic excited levels, the level with a certain value of J is expected to contribute to the relaxation process. It was proposed that, for Van Vleck paramagnets, the contribution of the magnetic dipole fluctuation in magnetic crystalline electric field (CEF) excited states can be expressed by [7, 8, 9]

$$\left(\frac{1}{T_1}\right)_{\mathrm{exc}} \propto \sum_{\Gamma_i} |\langle\Gamma_i|J_z|\Gamma_i\rangle|^2 \frac{\exp(-E_{\Gamma_i}/k_{\mathrm{B}}T)}{Z}, \tag{3}$$

where Γ_i is a CEF eigenstate, E_{Γ_i} is the CEF splitting between Γ_i and the nonmagnetic ground state, and Z is the single ion partition function. We applied eqn.(3) to J-multiplet levels. Since CEF parameters for $EuNi_2Si_2$ have not been reported, we considered only the first excited level $^7F_1$. Moreover, for simplicity, we assumed the degeneracy of $^7F_1$ triplet which is actually lifted by CEF. Then eqn.(3) can be rewritten as

$$\left(\frac{1}{T_1}\right)_{\mathrm{exc}} \propto \frac{\exp(-E_{J=1}/k_{\mathrm{B}}T)}{1 + 3\exp(-E_{J=1}/k_{\mathrm{B}}T)}. \tag{4}$$

Using eqn.(2) and (4), we performed the least-squares fit to the experimental $1/T_1$ and then obtained $E_{J=1}(=\lambda) \simeq$443 K. This value is comparable to other Eu^{3+} compounds, such as $EuPd_3$($\lambda \simeq$480 K) [3] and $EuBO_3$(λ=471 K) [10].

Next we discuss $K(T)$. In the Russell-Saunders coupling scheme, the temperature dependence of χ of Van Vleck paramagnets is known to be calculated when S, L, J, and λ are given [1, 3, 10]. In order to deduce $K(T)$ for $EuNi_2Si_2$, we employed the simplified model: the same value of

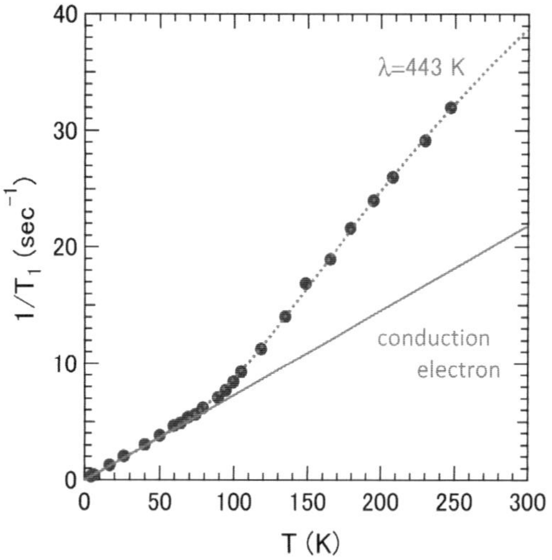

Figure 3. Temperature dependence of $1/T_1$. The solid green line represents the contribution of conduction electrons. The dotted red line represents the result of fitting by considering the contributions of conduction electrons and the $^7F_1$ multiplet.

the hyperfine coupling fields $H_{\rm hf}$ from the Curie law term and temperature independent term of each J level in χ. Then we applied the commonly used relation of

$$K(T) = K_0 + \frac{H_{\rm hf}}{N_{\rm A}\mu_{\rm B}}\chi, \tag{5}$$

where K_0, $N_{\rm A}$, and $\mu_{\rm B}$ are an on-site contribution of a conduction electrons, Avogadros number, and the Bohr magneton, respectively. Using eqn.(5), $K(T)$ above 100 K could be reproduced by the calculated χ with $H_{\rm hf} \simeq -0.81$ T/$\mu_{\rm B}$ and K_0=0.07% as drawn by the red line in Fig.2. At low temperatures, $K(T)$ continues to decrease with temperature in contrast to the temperature independent behavior of the calculated data. As shown in the inset of Fig. 2, we observed that the full width at half maximum (FWHM) of ^{29}Si NMR line divided by $|K(T)|$ showed an apparent increase below 100 K where $K(T)$ deviated from the calculation curve. These facts imply that $K(T)$ at low temperatures is affected by a certain amount of impurity including Eu^{2+} which bears a large magnetic moment. The CEF determination and the discussion about $H_{\rm hf}$ from each CEF level using the single crystal of $EuNi_2Si_2$ are challenges for the future.

5. Conclusion

We have performed ^{29}Si NMR measurement on the intermetallic Van Vleck paramagnet $EuNi_2Si_2$. The divalent Eu ions exhibit the unusual magnetic properties which arise because of its closely spaced J-muliplet levels above the nonmagnetic ground level. We observed the rapid increase in $1/T_1$ above 100 K which is explained by the fluctuation of magnetic dipole in the first excited state populated mainly at high temperatures. Based on the Russell-Saunders coupling scheme, $K(T)$ could be roughly reproduced by the deduced value of λ from $1/T_1$. Therefore, in the case of Eu^{3+}, higher J levels contribute appreciably to both $1/T_1$ and $K(T)$.

Acknowledgments

This work was supported by JPSJ KAKENHI Grant Number 26800193.

References

[1] Van Vleck J H 1932 *The Theory of Electric and Magnetic Susceptibilities (Oxford University Press, Oxford)* p.235
[2] Harris I R and Raynor G V 1965 *J. Less-common Met.* **9** 263
[3] Takeuchi T, Nakamura A, Hedo M, Nakama T, and Ōnuki Y 2014 *J. Phys. Soc. Jpn.* **83** 114001

[4] Bauminger E R, Felner I, and Ofer S 1978 *J. Magn. Magn. Mater.* **7** 317
[5] Seiro S, Kummer K, Vyalikh D, Caroca-Canales N, and Geibel C 2013 *Phys. Status Solidi B* **250** 621
[6] Wortmann G, Nowik I, Perscheid B, Kaindl G, and Felner I 1991 *Phys. Rev. B* **43** 5261
[7] Sugawara K 1977 *J. Phys. Soc. Jpn* **42** 1161
[8] Tanida H, Takagi S, Suzuki H S, Satoh I, and Komatsubara T 2006 *J. Phys. Soc. Jpn.* **75** 074721
[9] Tokunaga Y, Sakai H, Kambe S, Sakai A, Nakatsuji S, Harima H 2013 *Phys. Rev. B* **88** 085124
[10] Takikawa Y, Ebisu S, and Nagata S 2010 *J. Phys. Chem. Solids* **71** 1592

IOP Conf. Series: Journal of Physics: Conf. Series **868** (2017) 012024 doi:10.1088/1742-6596/868/1/012024

Synthesis and study of electronic state of $Sr_2CrO_2Co_2As_2$ with CoAs conduction layers

Atsushi Suzuki, Hiroto Ohta and Hiroko Aruga Katori

Department of Applied Physics, Graduate School of Engineering, Tokyo University of Agriculture and Technology, Tokyo 184-8588, Japan

E-mail: h-katori@cc.tuat.ac.jp

Abstract. We successfully synthesized a new member of compounds with the CoAs layer, $Sr_2CrO_2Co_2As_2$, and its partially substituted systems $Sr_2CrO_2(Tm_xCo_{1-x})_2As_2$ (Tm = Fe, Ni), and measured magnetization and electric resistivity of these polycrystalline compounds. As a result of magnetic measurement for $Sr_2CrO_2Co_2As_2$, magnetic moments of Co do not construct an itinerant electronic ferromagnetism unlike other compounds with the CoPn (Pn=P and As) layers. Both $Sr_2CrO_2(Tm_xCo_{1-x})_2As_2$ with Tm = Fe and Ni also do not show an itinerant electronic ferromagnetism down to 2 K. For each solid solution of $Sr_2CrO_2(Fe_xCo_{1-x})_2As_2$ with $x > 0.0$, ρ weakly increases with the decrease of T at low temperature region, indicating that the mixed occupancy of Cr and Fe within the conducting layers occurs in $Sr_2CrO_2(Fe_xCo_{1-x})_2As_2$. We conclude that the absence of ferromagnetism in $Sr_2CrO_2Co_2As_2$ is due to the self-electron-doping from Cr to the conduction bands and the attempt to recover the ferromagnetism by the hole-doping effect is prevented by the mixed occupancy of Cr and Fe in $Sr_2CrO_2(Fe_xCo_{1-x})_2As_2$ with $x > 0.0$. The result of our structural analysis supports that the disappearance of itinerant electronic ferromagnetism in $Sr_2CrO_2Co_2As_2$ is due to the self-electron-doping from Cr.

1. Introduction

Itinerant electronic magnetism is an old but still interesting theme in strongly correlated electronic systems. The system with both properties of itinerant electronic characteristics and low-dimensional characteristics is one of the interesting topics. The family of compounds with the CoPn conducting layers (abbreviated as CoPn family hereafter) has recently attracted attention as a candidate for such a magnetic system. Among the CoPn family, LaCoPnO with space group of $P4/nmm$, which have the same structure as iron-based superconductor LaFeAs(O,F) [1], have been studied by several groups [2, 3, 4]. In spite of a relatively large distance between the CoPn layers (~8.5 Å), these compounds show ferromagnetism below Curie temperature T_C being about 50 K. This result was revealed that $3d$ electrons of Co, which are delocalized in the CoPn layers, are responsible for magnetism in these compounds. However, the construction of ferromagnetic ordering at finite temperature in these compounds is an evidence of existence of magnetic interactions between the CoPn layers, meaning that these compounds are not two-dimensional magnetic materials in a strict sense. It is a good method to increase a distance between the CoPn layers to enhance the two-dimensional characteristics. However, Sr_2ScO_3CoAs with space group of $P4/nmm$ show itinerant electronic ferromagnetism with T_C=48 K, although the distance between the CoAs layers in them is about 16 Å, which is about twice of the distance between the CoAs layers in LaCoAsO [5]. On the other hand, the layered compounds AeCo$_2Pn_2$ (Ae

= Sr and Ba) with space group of $I4/mmm$ do not show any magnetic orderings, although the distance between the CoPn layers is about 6 Å[6, 7, 8, 9, 10]. To understand the origin of the magnetism of the CoPn family from the viewpoint of symmetricalness of the crystal structure, it is very important whether other compounds with the CoPn layers show a magnetic ordering. From this consideration, we focus on a new compound with the CoAs layers, $Sr_2CrO_2Co_2As_2$. By similarity with $Sr_2CrO_2Fe_2As_2$ [11], this compound is expected to have a body-centered tetragonal structure with space group of $I4/mmm$, and have a distance between the CoAs layers of about 9 Å. In this paper, we report the results of synthesis and magnetic and electric resistivity measurements of polycrystalline sample of $Sr_2CrO_2Co_2As_2$. We also report the properties of $Sr_2CrO_2(Tm_xCo_{1-x})_2As_2$ (Tm = Fe, Ni) to clarify the influence of the carrier doping into $Sr_2CrO_2Co_2As_2$ on its magnetism. We discuss the electronic state of $Sr_2CrO_2Co_2As_2$ on the basis of these experimental results.

2. Experiments

Polycrystalline samples of $Sr_2CrO_2Co_2As_2$ and $Sr_2CrO_2(Tm_xCo_{1-x})_2As_2$ (Tm = Fe, Ni) were synthesized through the following steps. First, CoAs and TmAs were synthesized by heating stoichiometric mixtures of Co (purity: 99.9%), Fe (99.9%) or Ni (99.9%) and As (99.9%) at 800 °C in evacuated silica tubes. Second, the obtained CoAs or TmAs powders were mixed with Sr (99%) and Cr_2O_3 (99.9%) in the appropriate ratio. These processes were performed in a glove box filled with highly pure N_2 gas to prevent the oxidation of Sr. The mixtures were sealed in evacuated silica tubes and heated at 1000 °C for 10 hours. After regrinding, the obtained samples were heated again at 1000 °C for 10 hours. $Sr_2CrO_2(Tm_xCo_{1-x})_2As_2$ were synthesized by firing the mixture of $Sr_2CrO_2Co_2As_2$ and $Sr_2CrO_2Tm_2As_2$ in the ratio of 1-x:x at 1000 °C in evacuated silica tubes.

The obtained samples were characterized by powder X-ray diffraction (XRD) measurements utilizing RIGAKU Smart-Lab with Cu K_α radiation. The measurement of magnetization (M) of samples was made by using a vibrating sample magnetometer ($MagLab^{\mathrm{VSM}}$, Oxford Instruments) or Magnetic Property Measurement System (MPMS, Quantum Design). The temperature dependence of M was measured in magnetic field (H) of 1 kOe or 10 kOe. Electric resistivity (ρ) was measured by a conventional dc four-probe method.

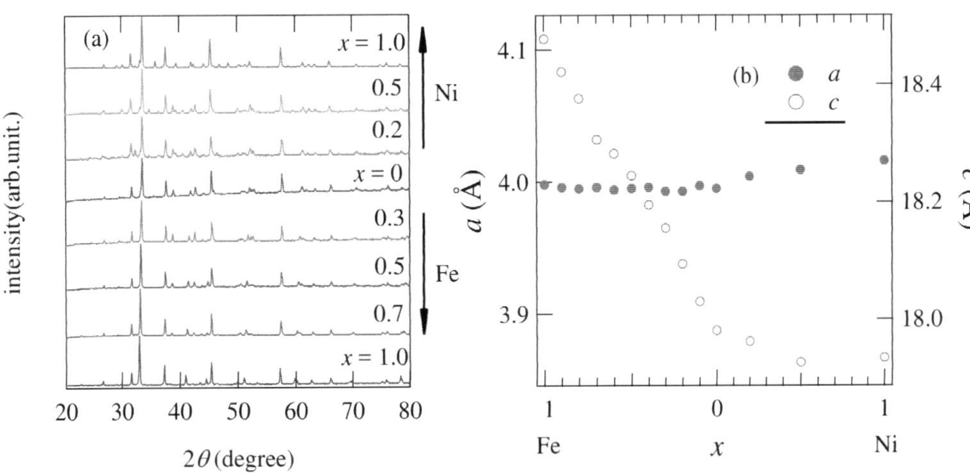

Figure 1. (a) Powder XRD pattern and (b) lattice parameters of $Sr_2CrO_2(Tm_xCo_{1-x})_2As_2$

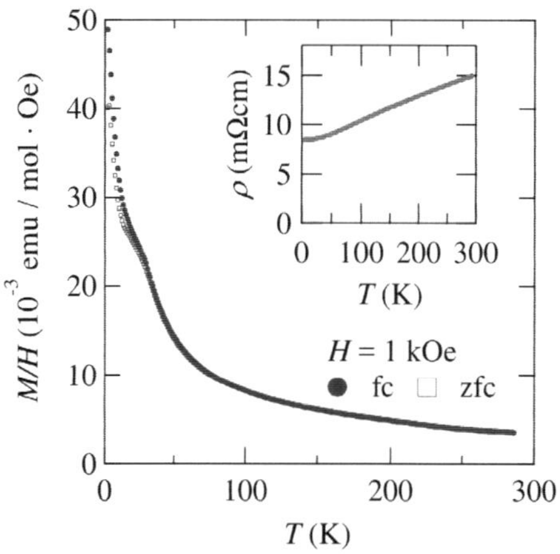

Figure 2. T dependence of M/H. Inset: T dependence of ρ.

3. Results and Discussion

Figure 1(a) shows the results of powder XRD measurement for $Sr_2CrO_2Co_2As_2$ and $Sr_2CrO_2(Tm_xCo_{1-x})_2As_2$. The powder pattern of $Sr_2CrO_2Co_2As_2$ could be fitted successfully with a body-centered tetragonal structure with space group of $I4/mmm$. Each peak position of $Sr_2CrO_2(Tm_xCo_{1-x})_2As_2$ changes systematically depending on x, indicating the success of synthesizing solid solution of $Sr_2CrO_2(Tm_xCo_{1-x})_2As_2$. Lattice parameters a and c of $Sr_2CrO_2(Tm_xCo_{1-x})_2As_2$ determined from XRD measurements are shown in Figure 1(b). For $Sr_2CrO_2(Fe_xCo_{1-x})_2As_2$, a is almost constant, while c rises with x. Both values of a and c obtained from the sample with x=1.0 we synthesized agree with the values already reported [11]. For $Sr_2CrO_2(Ni_xCo_{1-x})_2As_2$, a increases gradually, while c decreases with x.

Figure 2 shows temperature (T) dependence of M/H of $Sr_2CrO_2Co_2As_2$. The T dependence of ρ is also shown in the inset of this figure. The compound does not show any itinerant electronic ferromagnetic ordering down to 2 K, though T dependence of ρ exhibits a metallic character. The M/H shows an anomaly at about 35 K, below which the irreversibility was observed in M/H between zero-field cooling (zfc) and field-cooling (fc) measurements. Similar behavior has also been observed in $Sr_2MnO_2Zn_2As_2$ with non-conductive and non-magnetic ZnAs layer [12]. Therefore, we conclude that in $Sr_2CrO_2Co_2As_2$ the magnetic moment of Co with delocalized electrons does not construct an itinerant electronic ferromagnetism, which is quite different from the behavior observed in other CoPn family. The anomaly observed around 35 K may be caused by magnetic moment of localized Cr like Mn in $Sr_2MnO_2Zn_2As_2$.

To reveal why $Sr_2CrO_2Co_2As_2$ does not show an itinerant electronic ferromagnetism, we studied the electronic state of $Sr_2CrO_2Co_2As_2$ through the carrier doping. Usually, the substitution of Ni for Co corresponds to the doping of electrons to the conduction bands of CoPn family and that of Fe for Co does to the doping of holes [13]. Figure 3 shows T dependences of M/H of $Sr_2CrO_2(Tm_xCo_{1-x})_2As_2$. Each solid solution does not show an itinerant electronic ferromagnetism observed in other CoPn family. Anomalies caused by localized magnetic moment of Cr as observed in x = 0.0 were also observed in all of x in the both solid solutions. As the increase of x for Tm = Fe, the temperature the anomaly appears increases and reaches at about 280 K for $Sr_2CrO_2Fe_2As_2(x = 1)$ as reported in ref. [11]. While the temperature indicating the anomaly decrease with the increase of x for Tm = Ni. From these magnetic measurements of $Sr_2CrO_2(Tm_xCo_{1-x})_2As_2$, we can conclude that the carrier doping does not induce any ordering of magnetic moments of Co.

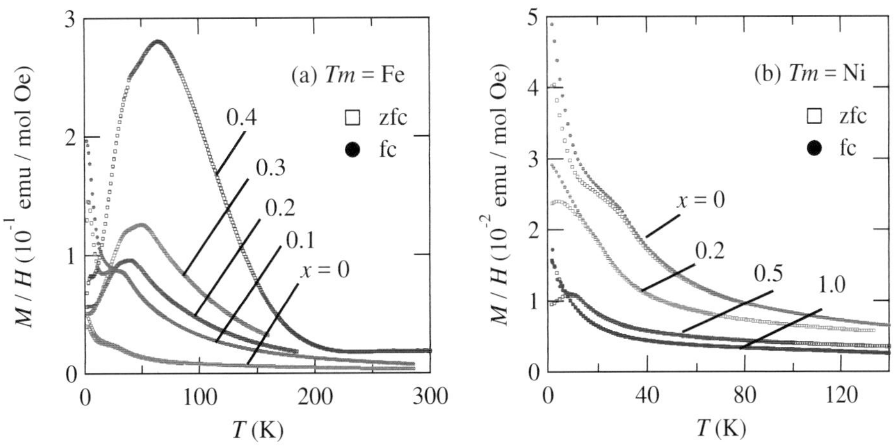

Figure 3. T dependences of M/H of $Sr_2CrO_2(Tm_xCo_{1-x})_2As_2$ ((a) Tm = Fe and (b) Tm = Ni).

The Sr_2VO_3FeAs shows a superconductivity without chemical modification or physical pressure unlike other iron-based superconductors. The appearance of superconductivity in Sr_2VO_3FeAs has been explained on the basis of several considerations. One of them is a self-electron-doping effect from V to the conduction bands in this system, which realizes a superconductivity. If similar self-electron-doping occurs in $Sr_2CrO_2Co_2As_2$, that is, the electron doping from Cr to the conduction bands occurs, its electronic state may have too much electrons to establish the itinerant electronic ferromagnetic ordering. In this case it is no wonder that itinerant electronic ferromagnetism is absent in $Sr_2CrO_2Co_2As_2$. For this situation the hole-doping to the conduction bands is usually considered to be an useful method to recover the ferromagnetism. However, ferromagnetic ordering due to Co moment does not occur in the hole-doping compounds of $Sr_2CrO_2(Fe_xCo_{1-x})_2As_2$.

We measured T dependence of ρ to clarify the cause that any ferromagnetism is not constructed in $Sr_2CrO_2(Fe_xCo_{1-x})_2As_2$. Figure 4 shows T dependence of ρ/ρ_{280K} of $Sr_2CrO_2(Fe_xCo_{1-x})_2As_2$, where ρ_{280K} is the resistivity observed at 280 K. The ρ of these samples are about 10 mΩcm in the whole temperature region below 280 K. For x=1.0, ρ shows semiconductor-like behavior, which is in agreement with behavior reported by Ref.[11]. On the other hand, for each solid solution with x <1.0, ρ shows metallic behavior at high temperature region, though ρ weakly increases with the decrease of T at low temperature region except for x=0.0. The degree of upturn at low temperature region develops gradually with the increase of x. Such an upturn behavior of ρ usually recognizes as the Kondo effect, and it has been also reported in $CaFe_{1-x}Cr_xAsF$, where Cr in the conducting layers acts as a scattering center for the itinerant electrons [14]. In contrast, the mixed occupancy of Cr and Fe in the conducting layers has been found in Sr_2CrO_3FeAs, which affects their electronic structure and induces the absence of superconductivity in this compound [15]. Based on these reports, we expect that the mixed occupancy of Cr and Fe in the conducting layers occurs in $Sr_2CrO_2(Fe_xCo_{1-x})_2As_2$ with 0.0< x ≤1.0 and it prevents the appearance of superconductivity, unlike the FeAs-layered compounds substituting Co for Fe, e.g. Ae(Fe,Co)$_2$As$_2$ (Ae = Sr, Ba) and Ca(Fe,Co)AsF [14, 16, 17]. Therefore, we can conclude that the disappearance of itinerant electronic ferromagnetism in $Sr_2CrO_2(Fe_xCo_{1-x})_2As_2$ except for x=0.0 is due to the mixed occupancy of Cr and Fe in the conducting layers, and the attempt to recover the ferromagnetism by tuning the Fermi energy up to peak position of density of states is prevented by this mixing

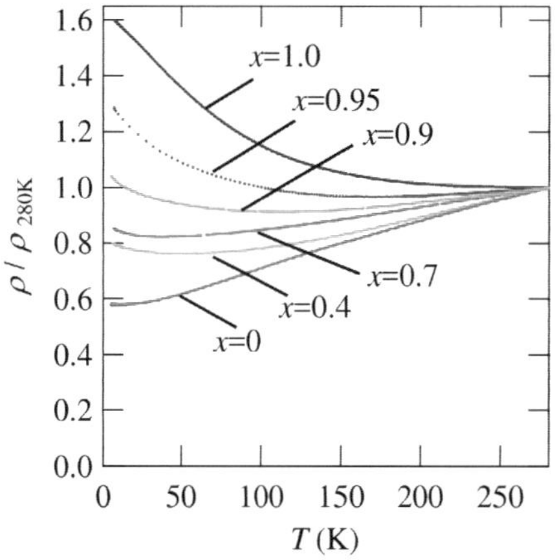

Figure 4. T dependences of $\rho/\rho_{280\text{K}}$ of $Sr_2CrO_2(Fe_xCo_{1-x})_2As_2$.

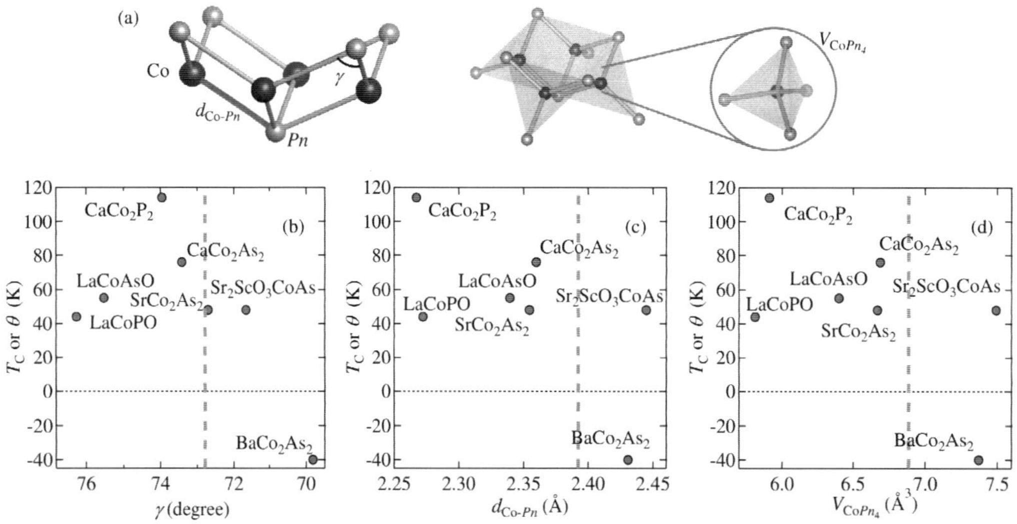

Figure 5. (a) Structural parameters of CoPn layer and CoPn_4 tetrahedron, γ, $d_{\text{Co}-Pn}$ and $V_{\text{Co}Pn_4}$. T_C or θ plotted against (b) γ, (c) $d_{\text{Co}-Pn}$ and (d) $V_{\text{Co}Pn_4}$. The blue dotted lines in (b), (c) and (d) indicate the location of the structural parameter of $Sr_2CrO_2Co_2As_2$. Data plotted in each figure are after Refs. [2, 4, 5, 13, 19, 20, 21, 22]

of Fe with Cr.

In the case of iron-based superconductors with the FeAs conducting layers, a certain correlation exists between a superconducting transition temperature and parameters express $FeAs_4$ tetrahedral structure [18]. Using the method like this, we sort out the relation between a transition temperature and parameters related to CoPn_4 tetrahedral structure for the CoPn family. Here we focus on three parameters, γ, $d_{\text{Co}-Pn}$ and $V_{\text{Co}Pn_4}$, which are the angle of Co-Pn-Co, the distance between Co and Pn and the volume of CoPn_4 tetrahedron shown in

Figure 5 (a), respectively. Figure 5(b), (c) and (d) show T_C or Curie-Weiss constant θ plotted against γ, d_{Co-Pn} or V_{CoPn_4}, respectively. As shown in Fig.5(b), (c) and (d), each parameter observed for $Sr_2CrO_2Co_2As_2$ has a value close to that observed for other CoPn family shown an itinerant electronic ferromagnetism. This result indicates that the occurrence of the phase transition to itinerant electronic ferromagnetism seems to be feasible in $Sr_2CrO_2Co_2As_2$ from the viewpoint of $CoAs_4$ tetrahedral structure. This structural analysis supports that the reason not to show the ferromagnetism is the self-electron-doping from Cr to the conduction bands in $Sr_2CrO_2Co_2As_2$.

4. Conclusion

We successfully synthesized a new polycrystalline compound with the CoAs layer, $Sr_2CrO_2Co_2As_2$, and revealed that this compound does not show any magnetic ordering of Co with delocalized electrons. The $Sr_2CrO_2(Tm_xCo_{1-x})_2As_2$ (Tm=Fe, Ni) also do not show any magnetic ordering of Co. The ρ for $Sr_2CrO_2(Fe_xCo_{1-x})_2As_2$ with x >0.0 show a temperature dependence to suggest the mixed occupancy of Cr and Fe in the conducting layers. From these experimental results, we conclude that disappearance of ferromagnetism is due to the self-electron-doping from Cr to the conduction bands in $Sr_2CrO_2Co_2As_2$ and the recovery of the ferromagnetism by the hole-doping due to the substitution of Fe for Co is prevented by the mixed occupancy of Cr and Fe in the conducting layers occurring in $Sr_2CrO_2(Fe_xCo_{1-x})_2As_2$. Our structural analysis also supports that the disappearance of itinerant electronic ferromagnetism in $Sr_2CrO_2Co_2As_2$ is due to the self-electron-doping from Cr.

Acknowledgments

This work was supported by Grants-in-Aid for Young Scientists (B) from Japan Society for the Promotion of Science (Grant Nos. 24760534 and 15K18211).

References

[1] Kamihara Y, Watanabe T, Hirano M and Hosono H 2008 *J. Am. Chem. Soc.* **130** 3296
[2] Yanagi H, Kawamura R, Kamiya T, Kamihara Y, Hirano M, Nakamura T, Osawa H and Hosono H 2008 *Phys. Rev.* B **77** 224431
[3] Sefat A S, Huq A, McGuire M A, Jin R, Sales B C, Mandrus D, Cranswick L M D, Stephens P W and Stone K H 2008 *Phys. Rev.* B **78** 104505
[4] Ohta H and Yoshimura K 2009 *Phys. Rev.* B **79** 184407
[5] Ohta H, Noguchi D, Nabetani K and Aruga Katori H 2013 *Phys. Rev.* B **88** 094441
[6] Imai M, Michioka C, Ohta H, Matsuo A, Kindo K, Ueda H and Yoshimura K 2014 *Phys. Rev.* B **90** 014407
[7] Jia S, Williams A J, Stephens P W and Cava R J 2009 *Phys. Rev.* B **80** 165107
[8] Pandey A, Quirinale D G, Jayasekara W, Sapkota A, Kim M G, Dhaka R S, Lee Y, Heitmann T W, Stephens P W, Ogloblichev V, Kreyssig A, McQueeney R J, Goldman A I, Kaminski A, Harmon B N, Furukawa Y, Johnston D C 2013 *Phys. Rev.* B **88** 014526
[9] Imai M, Ohta H, Michioka C, Ueda H and Yoshimura K 2012 *Proceedings of the 2012 Powder Metallurgy World Congress & Exhibition* 16F-T14-121
[10] Sefat A S, Singh D J, Jin R, McGuire M A, Sales B C and Mandrus D 2009 *Phys. Rev.* B **79** 024512
[11] Eguchi N, Ishikawa F, Kodama M, Wakabayashi T, Nakayama A, Ohmura A and Yamada Y 2013 *J. Phys. Soc. Jpn.* **82** 045002.
[12] Nath R, Garlea V O, Goldman A I and Johnston D C 2010 *Phys. Rev.* B **81** 224513
[13] Ohta H, Akabane E and Aruga Katori H 2015 *Phys. Proc.* **75** 309
[14] Matsuishi S, Inoue Y, Nomura T, Kamihara Y, Hirano M and Hosono H 2009 *New. J. Phys.* **11** 025012
[15] Tegel M, Hummel F, Su Y, Chatterji T, Brunelli M and Johrendt D 2010 *EPL* B **89** 37006
[16] Leithe-Jasper A, Schnelle W, Geibal C and Rosner H 2008 *Phys. Rev. Lett.* **101** 207004
[17] Sefat A S, Singh D J, Jin R, McGuire M A, Sales B C, Ronning F and Mandrus D 2009 *Physica* C **469** 350
[18] Lee C H, Iyo A, Eisaki H, Kito H, Fernandez-Diaz M T, Ito T, Kihou K, Matsuhata H, Braden M, and Yamada K 2008 *J. Phys. Soc. Jpn.* **77** 083704.
[19] Pfisterer M and Nagorsen G 1980 *Z. Naturforsch.* **35b** 703

[20] Cheng B, Hu B F, Yuan R H, Dong T, Fang A F, Chen Z G, Xu G, Shi Y G, Zheng P, Luo J L and Wang N L 2012 *Phys. Rev.* B **85** 144426
[21] Reehuis M, Jeitschko W, Kotzyba G, Zimmer B and Hu X 1998 *J. Alloys Comp.* **266** 54
[22] Pfisterer M and Nagorsen G 1983 *Z. Naturforsch.* **38b** 811

Effect of Co substitution on the magnetic properties in $Fe_{1-x}Co_xGa_3$

Yao Zhang[1], **Chishiro Michioka**[1], **Masaki Imai**[1], **Hiroaki Ueda**[1], **Kazuyoshi Yoshimura**[1,2]

[1] Department of Chemistry, Graduate School of Science, Kyoto University, Kyoto 606-8502, Japan
[2]Research Center for Low Temperature and Materials Sciences, Kyoto University, Kyoto 606-8502, Japan

E-mail: yao@kuchem.kyoto-u.ac.jp
kyhv@kuchem.kyoto-u.ac.jp

Abstract. We investigated the effect of Co substitution in $Fe_{1-x}Co_xGa_3$ by means of electric resistivity and magnetic measurements. Nearly ferromagnetic behaviors were observed in $Fe_{1-x}Co_xGa_3$ system by the chemical substitution of Co for Fe in diamagnetic mother compound $FeGa_3$, accompanying with a semiconducting to metallic transition. With the substitution of Co, the magnetization of $Fe_{1-x}Co_xGa_3$ increases first, and then decreases and finally it becomes diamagnetic. The magnetic properties of $Fe_{1-x}Co_xGa_3$ were analyzed in terms of self-consistent renormalization (SCR) theory and Takahashi's theory. The spin-fluctuation parameters, fourth order expansion coefficients of magnetic free energy ($\bar{F_1}$), the magnetic susceptibility in the ground state $\chi(0)$, the spectral widths of the spin fluctuation spectrum T_0 and T_A in the frequency and wave-number spaces, respectively, were estimated from the parameters derived from magnetic measurements. The magnetic properties of $Fe_{1-x}Co_xGa_3$ can be understood almost quantitatively by the SCR theory and Takahashi's theory of spin fluctuations.

1. Introduction

Various stoichiometric intermetallic compounds with unusual properties, such as magnetic, thermal, and strong correlated behaviors etc., have been discovered. Most of these compounds are found to be metallic. However there exist a few exceptions, such as semiconductors $RuAl_2$ [1], $RuGa_3$ [2], Fe_2VAl [3], and $FeGa_3$ [4], which draw great attention for prospective thermoelectric properties due to their complex band structures. The thermoelectric properties of these compounds are studied [5, 6, 7, 8] for such purpose, and high Seebeck coefficient values are achieved subsequently.

Among those intermetallic semiconductors, $FeGa_3$ has attracted much interest recently because of its intriguing magnetic behavior at finite temperature. $FeGa_3$ crystallizes into a tetragonal structure with space group of $P4_2/mnm$, and Fe atoms make pairs and form dimers within the ab plain in $FeGa_3$. The energy gap E_g of $FeGa_3$ was estimated to be around 0.3 to 0.5 eV by various methods, such as photoemission and inverse photoemission spectroscopic study [9], electronic resistivity measurements [4], magnetic measurement in the high-temperature region [10], which was confirmed by the band structure calculations based on the density functional theory [11, 12]. The energy gap is assumed to be formed by strong hybridization between

d electrons of Fe and p electrons of Ga. Yin *et al.* suggested the presence of the strong antiferromagnetic coupling between spins on Fe sites of Fe dimers leads to the formation of spin singlets [12]. Around 600 K, it shows the transition from semiconducting to a paramagnetic metal because of the thermally excited electrons above the energy gap [10]. Here we report paramagnetic properties of Co substituted $Fe_{1-x}Co_xGa_3$ systems based on the SCR theory and Takahashi's theory of spin fluctuations.

2. Experiment

Single crystals of $Fe_{1-x}Co_xGa_3$ were grown with Ga-self flux method. A mixture of highly pure Fe, Co, and Ga elements in ratio of Fe : Co : Ga = 1-X : X : 9 ($X \in [0, 1]$) were placed and sealed in quartz ampoules. The mixture was melted and homogenized in a furnace at 1273 K and then cooled down to room temperature slowly. Excess Ga flux was removed by decanting and corrosion of HCl acid. The obtained single crystals were annealed at 700 K for weeks for homogeneity. The purity of obtained crystals were checked by Cu-Kα_1 radiation, and lattice parameters were calculated using the Rietveld method. The chemical composition were measured by energy-dispersive x-ray spectroscopy (EDS). Magnetization curves were measured up to 7 T using a Quantum Design superconducting quantum interference device magnetometer. The electrical resistivity were measured by a home-built quadrupole electrical conductivity measuring apparatus from 5 to 300 K.

3. Results and Discussion

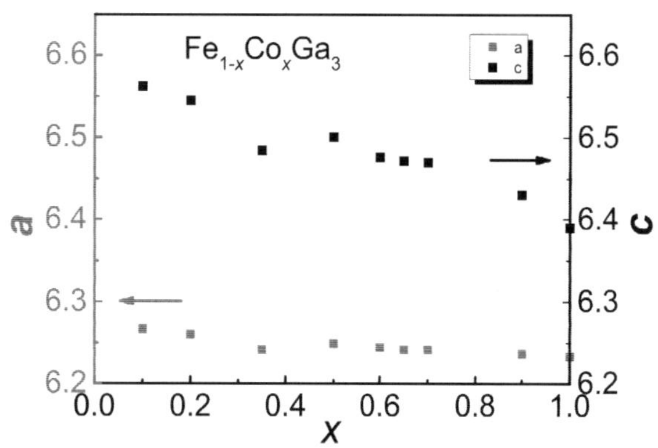

Figure 1. Lattice parameters a and c as functions of x for $Fe_{1-x}Co_xGa_3$ represented by red squares and black squares, respectively.

Structural characterizations at room temperature were carried out on powdered crystals of $Fe_{1-x}Co_xGa_3$ using XRD, and no evidence of impurity phase was found in various $Fe_xCo_{1-x}Ga_3$. The lattice parameters were estimated by the refined diffraction patterns and shown in Fig. 1. The values of a and c are in good agreement with former reports [13, 14]. The Co substitution in $FeGa_3$ leads to monotonic decreases in both lattice parameters a and c with increasing x, following Vegard's law.

Figures 2 and 3 show the temperature dependence of electric resistivity ρ for mother compound $FeGa_3$ and Co-substituted $Fe_xCo_{1-x}Ga_3$, respectively. The resistivity of $FeGa_3$ shows a Kondo-insulator like behavior. In the range of temperatures $5 < T < 180$ K, electrons in donor levels are excited to conduction band by thermal energy and resistivity decreases with increasing temperature. $FeGa_3$ begins to show metallic behavior from 180 to 260 K, suggesting the electron donors are thermally activated and saturated in this range. The resistivity decreases again with further increasing temperature above 260 to 300 K, and is shown as the inset (a) on a logarithmic scale. As shown in the inset (a), the electric resistivity above 260 K is linear in logarithmic scale, and is well described by the Arrhenius law $\rho(T) = \rho_0 \exp[(E_g/(2k_BT)]$, which gives energy gap

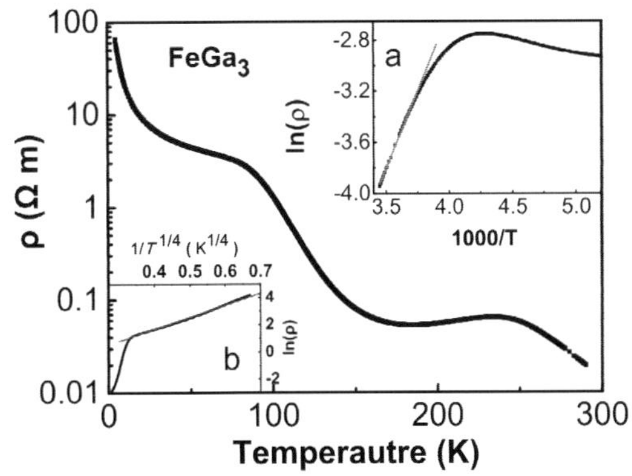

Figure 2. Electrical resistivity ρ for $FeGa_3$. The insets (a) and (b) show the resistivity on a logarithmic scale for $FeGa_3$ above 180 K, and that in the range of $5 < T < 180$ K respectively.

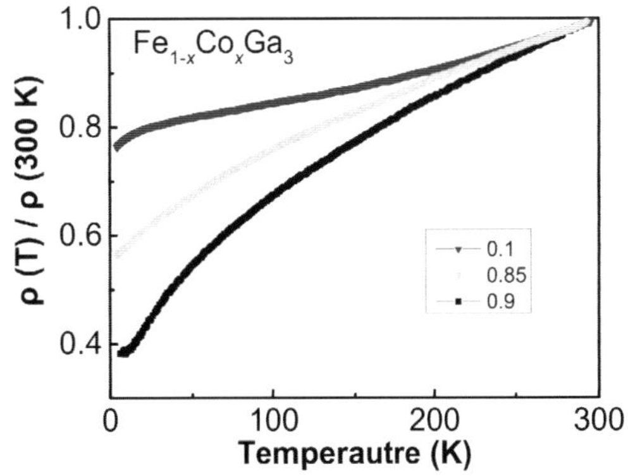

Figure 3. Normalized electric resistivity $\rho(T)/\rho(300K)$ of $Fe_{1-x}Co_xGa_3$ with $x = 0.1, 0.85$, and 0.9.

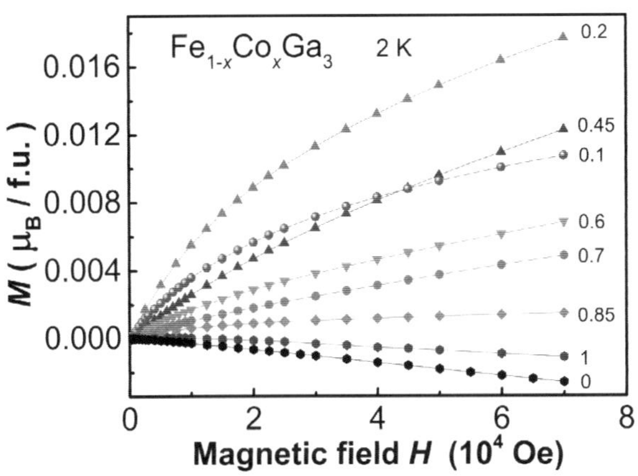

Figure 4. Isothermal magnetization curves of $Fe_{1-x}Co_xGa_3$ at 2 K with $x = 0, 0.1, 0.2, 0.45, 0.6, 0.7, 0.85$, and 1.

E_g of $FeGa_3$ approximately equals to 0.4 eV, in agreement with former reports. The inset (b) resistivity follows $\rho(T) \propto \exp[(T_M/T)^{1/4}]$ relation at low temperature, indicating the variable range hopping (VRH) among the Anderson localized states in 3D systems. Such dependence is also seen in few lightly doped semiconductors and Kondo-like insulators. Fig. 3 shows the effect of electron doping by Co substitution for Fe on normalized electric resistivity $\rho(T)/\rho(300K)$ of $Fe_xCo_{1-x}Ga_3$. The resistivity of $Fe_xCo_{1-x}Ga_3$ increases with increasing temperature, and the residual resistivity ratio is relatively small, suggesting $Fe_{1-x}Co_xGa_3$ belongs to a bad metal.

The magnetic ordering was checked by the magnetization measurement as a function of applied magnetic field at 2 K in Fig. 4. Both end compounds, $FeGa_3$ and $CoGa_3$, are diamagnetic. The magnetization first increases with Co substitution and achieves maximum around $x = 0.2$, and then it decreases. The value of magnetization is significantly small, and no rapid increment nor saturation up to 7 T was observed in Fig. 4, indicating the paramagnetic nature of $Fe_{1-x}Co_xGa_3$ ($0.1 \leq x \leq 0.7$). More clear evidence of no magnetic ordering can be gained from the Arrott plots of the magnetization curves of $Fe_{1-x}Co_xGa_3$ in Fig. 5 to Fig. 8 for various compositions. No intersection of the Arrott plot with the positive half of the M^2 axis at any temperature is regarded as a criterion of no spontaneous magnetic moment. The larger the intercept of $Fe_{1-x}Co_xGa_3$ obtained from Arrott plots at 2 K, the stronger magnetization is

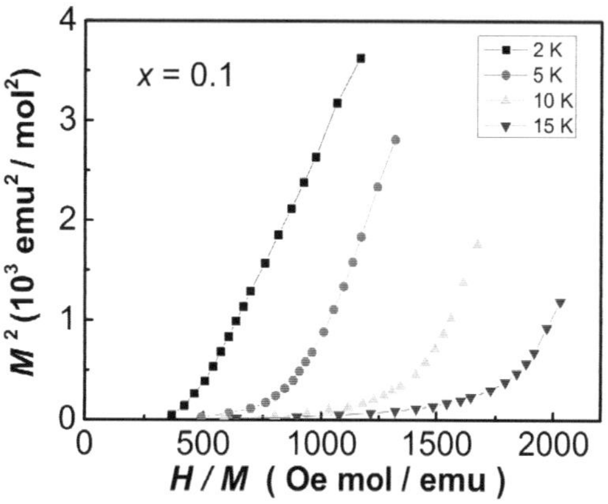

Figure 5. Arrott plots of $Fe_{1-x}Co_xGa_3$ with $x = 0.45$.

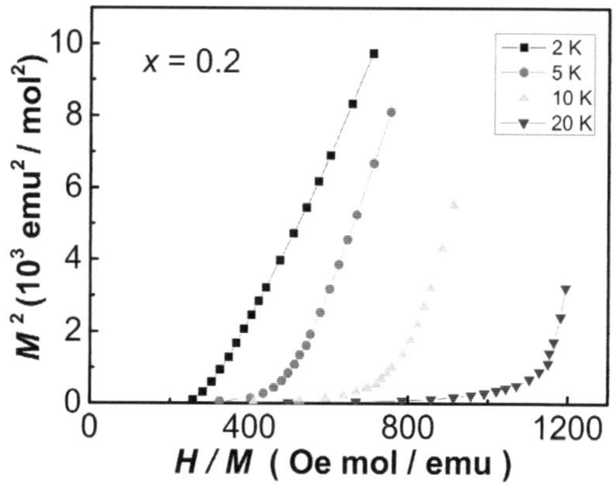

Figure 6. Arrott plots of $Fe_{1-x}Co_xGa_3$ with $x = 0.7$.

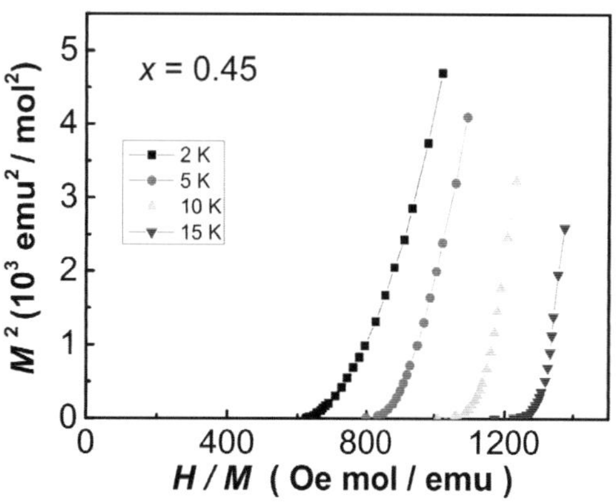

Figure 7. Arrott plots of $Fe_{1-x}Co_xGa_3$ with $x = 0.45$.

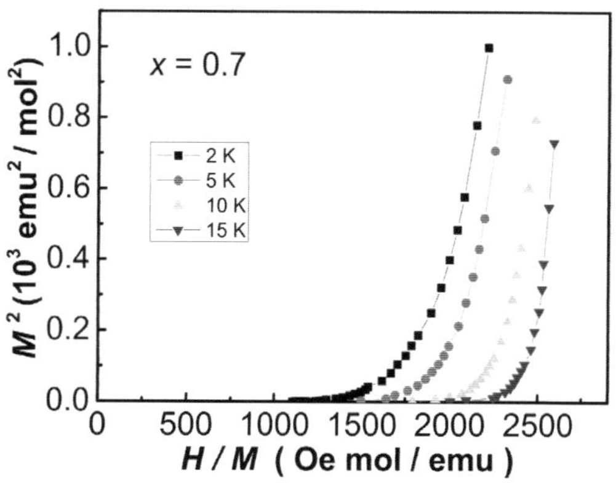

Figure 8. Arrott plots of $Fe_{1-x}Co_xGa_3$ with $x = 0.7$.

formed by the Co substitution.

Now we analyze the magnetic properties of $Fe_{1-x}Co_xGa_3$ according to self-consistent renormalizatoin (SCR) theory [15] and Takahashi's theory [16, 17] of spin fluctuations based on the idea of crucial roles of spin fluctuations on various magnetic properties in itinerant electron magnets. According to SCR theory, the imaginary part of dynamical magnetic susceptibility $\chi''(q,\omega)$ is characterized by two important parameters, T_0 and T_A, the measures of distribution widths of the frequency ω and wave-vector q spaces, respectively. In the SCR theory, another parameter, the mode-mode coupling constant $\bar{F_1}$ is regarded as the independent theoretical parameter. For exchange enhanced paramagnets, according to Takahashi, new parameter σ_p^2 corresponding to spontaneous magnetic moment squared in units of $(2\mu_{\rm B})^2$ is defined by ${\sigma_p}^2 = N_0/\chi(0)\bar{F_1}$. The temperature T_p corresponding to the Curie temperature for ferromagnets is also defined by $T_p = t_pT_0$, with t_p given by solving

$$A(0,t_p) = \frac{T_A}{15T_0}\sigma_p^2, \quad A(y,t) = \int_0^1 dz\,[\log u - 1/2u - \psi(u)]\,, \quad u = z(y+z^2)/t \tag{1}$$

where $y = N_0/\left[2T_A\chi(T)\right]$ and $A(y,t)$ is the reduced thermal spin fluctuation amplitude as a function of the reduced inverse of magnetic susceptibility y and reduced temperature $t = T/T_0$.

The digamma function is denoted by $\psi(u)$. With these parameters, σ_p and T_p, magnetic properties of paramagnets can be treated in the same way as ferromagnets.

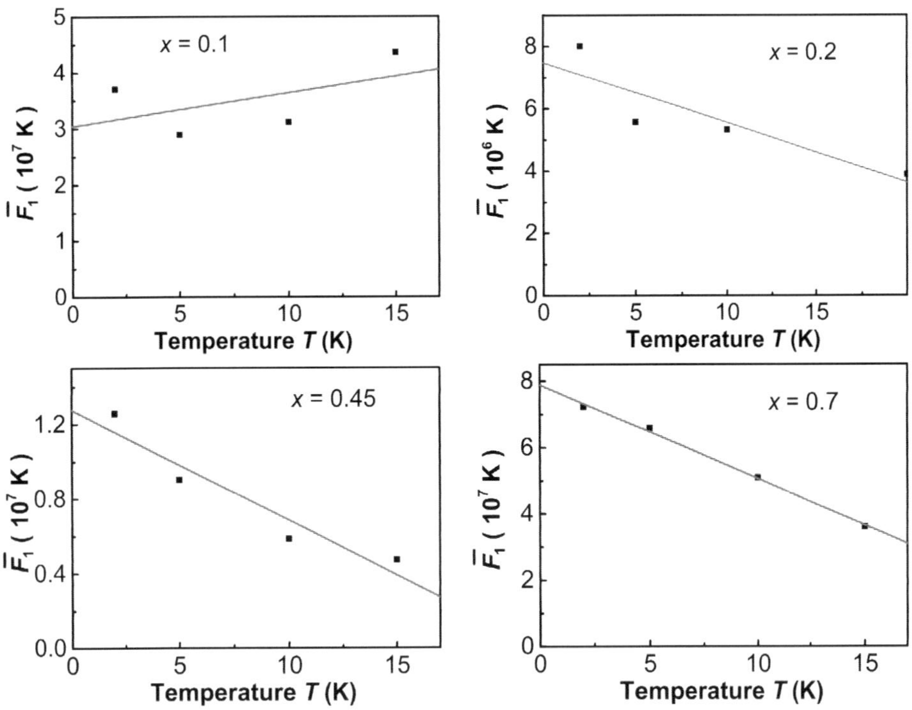

Figure 9. Estimation of $\bar{F_1}$ of $Fe_{1-x}Co_xGa_3$ with $x = 0.1, 0.2, 0.45$, and 0.7.

The parameter $\bar{F_1}$ represents the fourth expansion coefficient of the free energy in magnetization M in the ground state, the value of which can be estimated from the inverse slope of the Arrott plot of magnetization curve. In Fig. 9, values of $\bar{F_1}(T)$ estimated at several temperatures corresponding to those in Figures from 5 to 8 for $Fe_xCo_{1-x}Ga_3$ are plotted. The values of $\bar{F_1}$ in the ground state are estimated by extrapolating the linear fitting of those at finite temperatures to zero temperature. They are given by $3.04, 0.75, 1.28$, and 7.88 ($\times 10^7$ K) for $x = 0.1, 0.2, 0.45$, and 0.7, respectively. According to the theory of spin fluctuations [15, 16, 17], the following relation is satisfied among T_0, T_A, T_p and σ_p^2.

$$\sigma_p{}^2 = \frac{5T_0}{T_A} C_{4/3} \left(\frac{T_p}{T_0}\right)^{4/3}, \quad C_{4/3} = 1.00608\cdots. \tag{2}$$

By assuming the total spin-fluctuation amplitude is conserved (TAC) independently with temperature in Takahashi's theory, $\bar{F_1}$, T_A, and T_p follow:

$$\bar{F_1} = 2{T_A}^2/15cT_0. \tag{3}$$

where c is defined as the y linear coefficient of the zero-point fluctuation amplitude, given by $1/2$ for Lorentzian distribution function [17]. In addition, we can estimate the value of T_A from the T^2-linear coefficient of $1/\chi(T)$ at low temperature by using:

$$T_A = \frac{1}{\sigma_p^2}\left(\frac{75c}{8\alpha}\right)^{1/2}. \tag{4}$$

where α is defined by the relation: $\chi(0)/\chi(T) = 1 + \alpha T^2$. Then, the inverse magnetic susceptibility can be calculated using above parameters with the expression:

$$y = \frac{1}{c}[A(y,t) + A(0,t_p)]. \tag{5}$$

We calculated an inverse magnetic susceptibility for a typical composition $x = 0.45$. A comparison of the theoretical estimation versus experimental result for $Fe_{0.55}Co_{0.45}Ga_3$ is shown in Fig. 10. As shown in the figure, a tolerable agreement between experimental and calculated results are obtained, suggesting the importance of spin fluctuations in $Fe_{1-x}Co_xGa_3$.

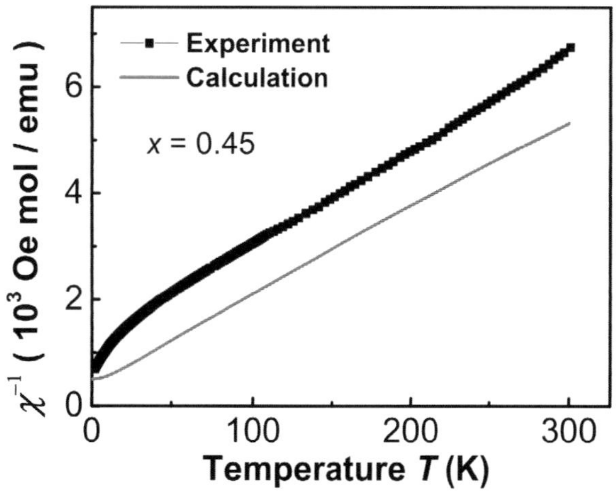

Figure 10. Temperature dependent inverse magnetic susceptibility of experimental measurement (black squares) and estimation based on theory (red line) for $Fe_{0.55}Co_{0.45}Ga_3$.

4. Summary

In summary, single crystals of $Fe_{1-x}Co_xGa_3$ with various compositions were successfully synthesized. Co substitution for Fe site drives drastic changes in $Fe_{1-x}Co_xGa_3$ on the electrical resistivity and the magnetic ground state. Transition from a diamagnetic mother host $FeGa_3$ to paramagnetic metals was observed. Magnetic properties of $Fe_{1-x}Co_xGa_3$ system are discussed in terms of the SCR theory and Takahashi's theory of spin fluctuations. The experimental and theoretical results achieved relatively good agreements with each other, suggesting the importance of spin fluctuations in $Fe_{1-x}Co_xGa_3$ system.

References

[1] Basov D N, Pierce F S, Volkov P, Poon S J and Timusk T 1994 *Phys. Rev. Lett.* **73**, 1865-1868
[2] Wagner-Reetz M, Cardoso-Gil R, Prots Y, Schnelle W and Grin Y 2014 *Solid State. Sci.* **32**, 56-60
[3] Nishino Y, Kato M, Asano S, Soda K, Hayasaki M and Mizutani U 1997 *Phys. Rev. Lett.* **79**, 1909-1912
[4] Hadano Y, Narazu S, Avila M A, Onimaru T and Takabatake T 2009 *J. Phys. Soc. Jpn.* **78**, 013702
[5] Mandrus D, Keppens V, Sales B C and Sarra J L 1998 *Phys. Rev.* B **58**, 3712-3716
[6] Amagai Y, Yamamoto A, Iida T and Takanashi Y 2004 *J. Appl. Phys* **96**, 5644-5648
[7] Takagiwa Y, Kitahara K, Matsubayashi Y and Kimura K 2012 *J. Appl. Phys* **12**, 123707
[8] Mikami M, Matsumoto A and Kobayashi K 2008 *J. Alloys Compd* **461**, 423-426
[9] Arita M *et al* 2001 *Phys. Rev.* B **83**, 245116
[10] Tsujii N *et al* 2008 *J. Phys. Soc. Jpn.* **77**, 024705
[11] Häussermann U, Boström M, Viklund P, Rapp O and Björnängen T 2002 *J. Solid State Chem.* **165**, 94-99
[12] Yin Z P and Pickett W E 2010 *Phys. Rev.* B **82**, 155202
[13] Umeo K, Hadano Y, Narazu S, Onimaru T, Avila M A and Takabatake T 2012 *Phys. Rev.* B **86**, 14442
[14] Kotur B, Babizhetskyy V, Bauer E, Kneidinger F, Danner A, Leber L and Michor H 2013 *Mater. Sci.* **49**, 211-219
[15] Moriya T 1985 *Spin Fluctuations in Itinerant Electron Magnetism* (Berlin Heiderberg: Springer)

[16] Takahashi Y 1994 *J. Phys.: Condens. Matter* **6** 7063-7073
[17] Takahashi Y 2013 *Spin Fluctuation Theory of Itinerant Electron Magnetism* (Heidelberg: Springer)